INDICATORS OF
ENVIRONMENTAL QUALITY

Environmental Science Research

Volume 1 — INDICATORS OF ENVIRONMENTAL QUALITY
Edited by William A. Thomas • 1972

INDICATORS OF ENVIRONMENTAL QUALITY

Proceedings of a symposium held during the AAAS meeting in
Philadelphia, Pennsylvania, December 26-31, 1971

Edited by
William A. Thomas

Group Leader, Environmental Indices
Environmental Program
Oak Ridge National Laboratory
Oak Ridge, Tennessee

℗ PLENUM PRESS · NEW YORK-LONDON · 1972

Library of Congress Catalog Card Number 72-86142
ISBN 0-306-36301-1

© 1972 Plenum Press, New York
A Division of Plenum Publishing Corporation
227 West 17th Street, New York, N. Y. 10011

United Kingdom edition published by Plenum Press, London
A Division of Plenum Publishing Company, Ltd.
Davis House (4th Floor), 8 Scrubs Lane, Harlesden,
London, NW10 6SE, England

Printed in the United States of America

"Our discussion will be adequate if it has as much clearness as the subject-matter admits of, for precision is not to be sought for alike in all discussions . . . [I]t is the mark of an educated man to look for precision in each class of things just so far as the nature of the subject admits . . . "

ARISTOTLE, 384–322 B.C.
Nicomachean Ethics
Book 1, Chapter 2

PREFACE

Researchers and agencies collect reams of objective data and authors publish volumes of subjective prose in attempts to explain what is meant by environmental quality. Still, we have no universally recognized methods for combining our quantitative measures with our qualitative concepts of environment. Not all of our environmental goals should be reduced to mere numbers, but many of them can be; and without these quantitative terms, we have no way of defining our present position nor of selecting positions we wish to attain on any logically established scale of environmental values. Stated simply, in our zeal to measure our environment we often forget that masses of numbers describing a system are insufficient to understand it or to be used in selecting goals and priorities for expending our economic and human resources.

Attempts at quantitatively describing environmental quality, rather than merely measuring different environmental variables, are relatively recent. This condensing of data into the optimum number of terms with maximum information content is a truly interdisciplinary challenge. When Oak Ridge National Laboratory initiated its Environmental Program in early 1970 under a grant from the National Science Foundation, the usefulness of environmental indicators in assessing the effects of technology was included as one of the initial areas for investigation. James L. Liverman, through his encouragement and firm belief that these indicators are indispensable if we are to resolve our complex environmental problems, deserves much of the credit for the publication of this book.

Unfortunately, the scientific literature is not abstracted or indexed with appropriate keywords that allow rapid dissemination of these new ideas and methodologies. To encourage cooperation among persons with diverse professional interests and to review the state of the art, the American Association for the Advancement of Science sponsored a two-day symposium entitled "Indicators of Environmental Quality" during its annual meeting in Philadelphia from December 26 to 31, 1971. The authors of the first 18 of the 21 chapters of this book prepared their papers for presentation at that symposium. The concluding three chapters were added to broaden even more the book's coverage of environment. Due to time limitations, not to a restricted viewpoint of what constitutes our environment, we had to restrict the symposium's scope primarily to the physical, chemical, and biological aspects. The social consequences of our actions are recognized throughout the volume, and one author directly addresses the social aspects of environment.

The symposium emphasized the need for public participation in decisions concerning environmental quality, and all speakers agreed that indicators facilitate the required communication among public officials, scientists, and the public. A total of 86 members of the audience submitted written questions for consideration during the panel discussions that followed each half-day session. All of the speakers had the opportunity of responding to the applicable questions before submitting the final revisions of their manuscripts. Speakers alone do not make a symposium successful, and the active participation of the audience contributed greatly to the success of this one. Special acknowledgement is due the individual authors who cooperated in many ways; their promptness in meeting the necessary deadlines ensures a timely volume.

Designing and testing indicators of environmental quality are not mere academic exercises — scientists have a responsibility to make "environment" comprehensible to all segments of society that justifiably demand a greater participatory role in determining the habitability of our planet. It is in the spirit of interdisciplinary action toward providing objective measures of environmental quality that this volume is published.

W.A.T

CONTENTS

INDICATORS OF ENVIRONMENTAL QUALITY:
AN OVERVIEW*

William A. Thomas

Leader, Environmental Indices Group
Environmental Program, Oak Ridge National Laboratory†
Oak Ridge, Tennessee 37830

A concerted effort to enhance habitability of our planet is unlikely to succeed unless we know "where we are" and "where we want to go." To answer these questions, we first must consider exactly what we include in the term "environment." If we restrict our consideration to overly simplified definitions, such as the amount of a specified pollutant in air, we have very little difficulty in measuring environment. However, as we broaden our definition to include all the physical components, or all the physical and biological, or all the physical, biological, and cultural ones, environment becomes exponentially more difficult to describe. Nonetheless, I believe we must take the holistic approach at the outset and define environment as that complex of interacting physical and cultural factors which routinely influences the lives of individuals and communities. This indeed is a broad definition, but we should not forget when we study the individual components that the entirety functions as a system of interacting components.

A major difficulty in describing environment is that all of its components cannot be measured directly. The challenging field of social indicators[2,3,9,10] provides ample evidence of the problems involved in assigning values to the social or cultural aspects of environment such as education, public order, and recreational opportunities. However, we can measure certain variables that indicate the presence or condition of phenomena that cannot be measured directly. These indicators reflect the state of any aspect or component of the environment. The method of selecting the indicator varies with the characteristics of the component, but they all share one requirement. The indicator must respond to changes in the component it is scaling in such a manner that it accurately reflects the magnitude of those changes. Objective aspects such as meteorological conditions seldom require an indicator because they can be measured directly, but indicators usually are necessary to assign quantitative values to subjective aspects such as public health.

Several indicators may be integrated into one index for the more complex components if data are available. An index is a composite value for an

*Research sponsored by the National Science Foundation RANN Program.

†Operated by Union Carbide Corporation for the U.S. Atomic Energy Commission.

1

environmental component for which we have more than one indicator; this distinction in terms is not accepted universally, but it is helpful in understanding the hierarchical nature of our environment. Ideally, these indices also can be aggregated by regions to establish geographical hierarchies. For example, we lack a universal measure for water quality even though we do measure the principal contaminants and thus have a series of indicators that can be combined into a single water quality index. Assignment of quantitative weighting factors to the values of the individual indicators poses complex questions due to incomplete scientific conclusions on the relative detrimental effects of some contaminants. Even the identification of suitable criteria for measurement is difficult for some pollutants; the undesirable effects of noise and odors, for example, are not easily characterized.

Because measures of the more subjective components of environmental quality, such as urban sprawl, uniqueness of open spaces, and scenic landscapes, cannot be devised as objectively as other measures might be, indices for them may incorporate value judgments by necessity. Obviously, the preferences even of prudent men often differ; therefore, use of personal judgments is minimized wherever possible.

Living organisms provide convenient full-time monitors of all pollutants, including their synergistic effects. Thus biological indicators, like the miners' canary, measure the actual responses of organisms or populations to environmental quality rather than predict a biological response from physical measurements. The physiological and ecological diversity of species allows a wide choice of indicator species for various environmental factors and situations. At the other end of the biological spectrum, biochemical reactions might possibly be used in tests for more specific classes of contaminants. Because ability to support life is a prime characteristic of any environment, the general vigor of natural populations provides a readily accessible gage of habitability which will be used more frequently as our ability to interpret population fluctuations increases. Systematic use of a series of biological indicators permits a more detailed description of quality.

We need to facilitate communication among the segments of society now concerned with environmental quality by providing adequate information in a compact format. The major goal in development of indicators is a translation by a scientifically defensible method of the many components of environment into an optimum number of terms with maximum information content. To do this, we accept some reduction in precision, but in return we gain the ability to communicate. We can classify the potential users of this information into five not-mutually-exclusive categories.

1. *Citizenry*. The majority of citizens realize that the quality of their environment is threatened, but the complexity of the environment precludes their understanding it. Indicators can enhance public sensitivity to environmental issues and encourage accountability of elected officials.

2. *Local and Regional Planners and Government Officials.* Alternatives for many decisions which affect our environment seldom are explored; again, complexity of the system discourages the attempt. Causally related indicators can provide predictions of consequences that may be expected from alternative courses of action. Objective allocation of public funds at all levels of government can be encouraged by the use of indicators to establish priorities for funding and to measure the effects of budgetary decisions.

3. *Judiciary, Legislature, and Regulatory Agencies.* These groups often attack environmental problems by establishment and enforcement of standards. A rational series of indicators is the precursor of a rational series of standards. The National Environmental Policy Act of 1969 requires the President to submit an annual report to Congress in which he shall convey the "status and condition of the major natural, manmade, or altered environmental classes of the nation" and the "current and foreseeable trends in the quality ... of such environments." Prose alone soon will be inadequate, and the Council on Environmental Quality now includes environmental indicators and indices in these reports.[4] Regulatory mechanisms for maintenance and enhancement of our natural and manmade resources often lack clear guidelines that allow intelligent appraisal of the effectiveness of our stewardship policies. Similarly, our courts will benefit from a system of objective indicators that quantitatively describe the impairment of resource quality by pollutants. Indices can link scientific knowledge with legal standards for environmental quality.

4. *Scientists and Engineers.* Research priorities for private and public resources can be set and defended with indicators in circumstances where perspective alone might fail. Indices provide a convenient format for summarizing and handling data and for presenting research results and proposals for action to responsible authorities, who also would find them useful in making their decisions.

5. *Special Interest Groups.* Many groups — including labor and industrial organizations, lobbies, public interest firms, conservation societies, and others — advocate positions on what constitutes desirable environmental quality and thus need quantitative terms to support those positions. By standardizing the use of data, indices can make the efforts of these groups more efficient to the benefit of all concerned.

No user group is likely to limit its use of environmental information to indicators and indices alone; in fact, data might not assist at all in resolving some problems. Certainly, all appropriate data will be examined when specific issues arise, but indices and indicators do serve as a convenient shorthand system that users can understand, appreciate, and utilize in attaining their objectives. Ample evidence and recommendations exist to indicate a genuine need for these indicators.[1,4-8] These indicators and indices are susceptible to misuse just as all information systems are, but I think they actually promote open discussion and retard the misleading uses of environmental information that may occur when only selected raw data are available to a limited number of individuals. One of the constituents of a successful indicator system must be the means of explaining its proper use and limitations to its users.

The available indicators today are useful in depicting trends through time and in comparing environmental quality among geographical areas. It obviously would be much more helpful to learn in advance what the expected environmental conditions will be than to be informed after the fact. This current limitation on the use of indicators results from the present data collection methods and is not an inherent fault of the indicator concept. We need to increase the predictive capability of indicators to permit prophylactic action when necessary. Sufficient data might not be available now to do this, but by coordination of development of techniques with expansion in the data collection system, many inefficiencies can be avoided. Rather than design indices to use the available data, as we now do, we should collect data that will be used most advantageously in the indices. In this manner we can learn from the indices themselves which data should be added or deleted from the system.

In addition, we should scale all our indicators to a common denominator, such as human health, so that we can discuss as many environmental variables as possible with a standard reference point. We then would know which problems should be given increased attention. Not all aspects of environment are compatible with a common denominator approach, and synergistic effects will remain troublesome, but the result nonetheless would be an improvement over our present efforts.

Development and communication of environmental indicators represents a truly interdisciplinary challenge that requires the merging of talents from many diverse disciplines and viewpoints. Perhaps Edna St. Vincent Millay best described our purpose when she wrote in her Sonnet Number CXL:

> Upon this gifted age, in its dark hour
> Rains from the sky a meteoric shower
> Of facts ——; they lie unquestioned, uncombined,
> Wisdom enough to leech us of our ill
> Is daily spun, but there exists no *loom*
> To weave it into fabric. (Italics added)

We should recognize the difficulties that must be overcome to produce this "loom" and should not hope for instant panaceas; but when we succeed, the beneficial consequences will be far ranging – objectivity will replace subjective rhetoric in assessment of shifts in environmental quality. Indeed, we will be able not only to define our goals but also to measure how well we progress toward them.

REFERENCES

References to detailed literature concerning the specific aspects of environment discussed here are cited in subsequent chapters.

1. American Chemical Society. *Cleaning Our Environment. The Chemical Basis for Action.* American Chemical Society, Washington, D.C., 1969.
2. Bauer, R. A. (ed.). *Social Indicators.* M.I.T. Press, Cambridge, 1966.
3. Bonjean, C. M., R. J. Hill, and S. D. McLemore. *Sociological Measurement. An Inventory of Scales and Indices.* Chandler Publishing Co., San Francisco, 1967.
4. Council on Environmental Quality. *Environmental Quality. The First Annual Report of the Council on Environmental Quality.* U.S. Government Printing Office, Washington, D.C., 1970. Subsequent annual reports also reflect the need for additional environmental indicators.
5. Environmental Pollution Panel, President's Science Advisory Committee. *Restoring the Quality of Our Environment.* U.S. Government Printing Office, Washington, D.C., 1965.
6. Environmental Study Group. *Institutions for Effective Management of the Environment.* National Academy of Sciences, Washington, D.C., 1970.
7. Fisher, J. L. The Natural Environment, *Ann. Amer. Acad. Political and Social Sci.* 371: 127–140 (1967).
8. Panel on Systematics and Taxonomy, Federal Council for Science and Technology. *Systematic Biology. A Survey of Federal Programs and Needs.* Office of Science and Technology, Washington, D.C., 1969.
9. Sheldon, E. B., and W. E. Moore (eds.). *Indicators of Social Change. Concepts and Measurements.* Russell Sage Foundation, New York, 1968.
10. U.S. Department of Health, Education, and Welfare. *Toward a Social Report.* U.S. Government Printing Office, Washington, D.C., 1969.

WHY ENVIRONMENTAL QUALITY INDICES?

Thomas L. Kimball

Executive Vice-President
National Wildlife Federation
1412 Sixteenth Street, NW
Washington, D.C. 20036

Science and technology in general, and professionalism in particular, are under attack in our great nation. The average American citizen is confused and confounded by purported scientific rationales that mix philosophy, politics, and opportunism with the analysis of factual research data. That confusion turns to disgust, and then rejection, when the public listens to professionals trained in our best institutions of higher learning and with similar long-term experience who come forth with diametrically opposed conclusions and recommendations from an analysis of environmental research data. Unfortunately, there appears to be a growing conviction among the American populace that it is extremely difficult if not impossible to get the unvarnished truth from the scientific community. As a consequence, professionalism is being downgraded in direct proportion to the inability of the public to guess which scientist has the right answer. Nowhere is there a better example of this phenomenon and dilemma than in an analysis of the current environmental–ecological revolution. In February, 1969, the National Wildlife Federation commissioned the Gallup Organization to survey public opinion on environmental degradation.[1] Gallup found that the majority of Americans were either somewhat or deeply concerned about the deteriorating quality of living brought about by the fouling of our own nests. Both fortunately and unfortunately, many and varied segments of our modern complex society have also become aware of this burgeoning public interest. It has brought forth new pollution abatement laws replete with political tub thumping, a new dimension to more intense and centralized government regulation, an alarmed industrial complex – insecure in its continued ability to pollute at the expense of the environment, the Madison Avenue hucksterings of endless products reportedly harmless to the plant and animal ecosystems, and a citizenry that is incapable of separating advertising fantasyland from the truth. Extremist "eco-freaks" and a rash of instant ecologists whose dire predictions have failed to materialize have created a king-size environmental credibility gap that threatens to destroy any significant progress toward the real solution of our critical environmental problems.

Government leadership has been a great disappointment. Whenever government finds itself in the middle of two aggressive constituencies with conflicting viewpoints, its reflexes are predictable. Instead of concentrating on an aggressive action program based upon the best scientific evidence available, government usually embraces one of two cop-outs: (1) a proposal for government reorganization is drafted or (2) a commission is appointed for further study. Both of these inept measures usually take years to complete, and neither effort has ever made any great contribution to the solution of a problem. Then there is a Congressional propensity to place in the same federal executive agency the mandate to both promote and regulate a specific important natural resource. The best examples are: the Atomic Energy Commission, which possesses the authority both to promote and to regulate the peaceful uses of atomic energy, and the Federal Power Commission, with authority to promote and regulate the energy needs of our nation. The Commerce Department is industry's spokesman in governmental affairs. Agriculture promotes and regulates the farmer and the commodities he produces. The entire executive branch of our federal government (and also to a lesser degree in state government only because of size) is rapidly losing the confidence of increasingly larger segments of our society because of its apparent willingness to react to principal special interest lobbies while unable or unwilling to respond to majority public opinion or to take the responsive actions essential to resolving critical issues.

The seniority system in our Congress perpetuates rules which essentially place in the hands of an individually powerful committee chairman the authority to prevent the entire Congress from considering, debating, and voting on critically needed legislation.

By now everyone should be convinced why we need Environmental Quality Indices. The lay citizen is interested in preserving and perpetuating a quality life in a natural environment in addition to keeping our nation strong, healthy, and secure from our enemies, both foreign and domestic, by preventing the misuse and waste of our great wealth of resources. The effort is now entering a crucial stage. The conservation, environmental, and ecological forces have been and are anxious to continue the battles which shape our national priorities. The man on the street must be armed with the facts and figures he considers accurate and has confidence in if we expect him to enter the fray with any significant firepower. The NWF's Environmental Quality (EQ) Index[2,3,4] is an effort designed to provide the concerned citizen with a comprehensive review of published information on factors affecting environmental quality presented in rather simple language and graphics readily understood by the masses.

While we are very proud of the innovative work done within the ranks of the National Wildlife Federation staff, we are not so foolish or short-sighted as to think that our EQ Index — now in its third year — represents the ultimate product or even the best analysis of available data. However, we do feel that it gives the average citizen a much better grasp of the environmental situation as it exists today and as it might look tomorrow and next year and in the foreseeable future.

The NWF EQ Index is the end product of an exhaustive, scholarly exercise that attempts to reduce reams of information — much of it disjointed at best and

some possibly erroneous at worst – into a simple, orderly, graphical representation of environmental conditions. We are persuaded that our EQ Index enables
the average reader to quickly grasp the overall environmental situation and to
"zero in" on the key issues. Armed with the information contained in a
comprehensive Environmental Quality Index, and the trends that it discloses, a
citizen is in a position to go through the accepted democratic processes of
examining the pros and cons of the issue, listening to the comments and advice
of others, and eventually reflecting his views in the various forums and forces
that shape our national policy.

For those of you who might not be familiar with the EQ Index prepared by
the National Wildlife Federation, I will spend a few minutes outlining its major
points and describing the processes by which some of the judgments are made
and by which some of the costs are determined.

The National Wildlife Federation's third annual Environmental Quality
Index was published in October. When first published in the fall in 1969, the EQ
Index evaluated six natural resources: air, water, soil, forests, wildlife, and
minerals. In 1970, a seventh item – living space – was added to the list.

The various categories of the environment were subjectively rated from best
to worst and then were scored on a numerical scale of 0 to 100, as listed in Table
1. A "0" would equal death or disaster; "100" would be ideal conditions with
environmental equilibrium. For example, soil is, in our relative judgment, in the
best condition of any single resource, but soil conditions are still far less than
ideal. In 1970, it was given an Index value of 80. This year, because of
continuing losses, the rating slipped to 78. Air is the natural resource that is in
the worst shape. It actually poses a danger to human health in many cities. The
1970 Index was placed at 35. Continuing pollution reduced this to 34 in 1971.

Table 1. Environmental categories and ratings

Category	1971 Score
Soil	78
Timber	76
Living space	58
Wildlife	53
Minerals	48
Water	40
Air	34

Next, the seven elements were assigned a relative importance value expressed
as a percentage. Some elements such as air, water, living space, and soil for food
are essential to life and must necessarily be assigned a higher value. Table 2
shows our evaluation of relative importance.

Table 2. Environmental categories and relative importance

Category	Relative importance
Soil	30
Air	20
Water	20
Living space	12.5
Minerals	7.5
Wildlife	5
Timber	5
	100%

Table 3. Development of national EQ index

Category	1971 Score	Relative importance	EQ Points
Soil	78	30	23.4
Air	34	20	6.8
Water	40	20	8.0
Living space	58	12.5	7.25
Minerals	48	7.5	3.6
Wildlife	53	5	2.65
Timber	76	5	3.80
		NWF National EQ Index:	55.50

To develop the EQ points for the comprehensive, overall index, the 1971 score for each category is applied to this percentage figure. Table 3 shows how our 55.5 National EQ Index was developed.

Quite obviously, a number of value judgments were arbitrarily made. These, in turn, led to the conclusions presented in the EQ Index. Our judgments and conclusions were necessarily subjective, but while we consider these judgments scrupulously fair, they are subject to challenge. In fact, we are hopeful the scientific community can help in eliminating those judgments which have insufficient back-up data by filling in the gaps in our body of knowledge about the biosphere in which we live.

Since trends are the important things, I think you would be interested in knowing what they are in the various environmental categories. You will note that only in the categories of water and timber have we made any gains since last year.

Soil quality has dropped from 80 to 78 with haphazard land development. Timber quality has risen slightly from 75 to 76 as growth exceeds cut. Living space quality has fallen to 58 from 60 as more people crowd into less space with more pollution. Wildlife quality has dropped to 53 from 55 as more and more habitat is converted to those uses adverse to wildlife. Minerals have slipped from 50 to 48 as we use up minerals and fossil fuels faster than we find them. Water quality is holding steady at the intolerably low level of 40 as we continue to develop tougher legislation but expend capital for physical plants only to the level of preventing further deterioration of water quality. Air quality — our most serious problem — continued to drop from 35 to the shockingly low level of 34 with more autos and more industrial pollutants. The overall trend shows a decline from 57 in 1970 to 55.5 in 1971.

The costs of pollution with respect to health, vegetation, materials, esthetics, and property values must be highlighted in any worthwhile EQ Index, because eventually the entire decision making process evolves to the point where the cost of pollution versus the cost of pollution control or abatement must be addressed head on. Before reasonable decisions can be made, the public must have an understanding — or at least an appreciation — of what the costs are of abating or controlling the various forms of pollution, and more importantly of what are the benefits both in economics and esthetics that will accrue to the individual if pollution is controlled. Armed with such factual data — even if it has been oversimplified and generalized by being incorporated in an EQ Index — the average citizen is in a position to vote more intelligently on pollution issues.

The average annual cost of air pollution in the United States ($16.1 billion) has a clear impact on the average taxpayer's purse (Table 4). Table 5 brings together a wealth of water pollution facts and figures in a forceful presentation of the price tag ($42.3 billion) attached to clean water.

We rely heavily on governmental agency reports, press conferences, and statements made by key government officials before Congressional committees and public groups to arrive at our rankings, individual scores, and relative importance. As part of our Annual EQ Index, we now publish a "Reference Guide" which is keyed directly to statements made in the EQ Index.[5] By checking the Reference Guide, one can quickly determine the reports, tabular data, and statements that contributed significantly to the final form taken by each of the seven environmental factors. A great many of our statistics are obtained from the Environmental Protection Agency and the Council on Environmental Quality. In particular, the CEQ's Annual Report on Environmental Quality is probably our single most important source of factual data.

Table 4. Dirty air costs

Source of cost	Total cost (10^9)	Cost for average family ($)
Cost to human health	6.1	117
Cost to residential property	5.2	100
Cost to materials	4.7	90
Cost to vegetation	0.1	2
Total	16.1	309

Table 5. The price tag for clean water

Source of improvement	Cost (10^9)
Municipal waste treatment plant construction costs	
Primary and secondary treatment	8.7
Tertiary treatment	3.9
Operation and maintenance	4.5
Industrial abatement costs	
Nonthermal	3.2
Reduce thermal pollution	2.0
Operation and maintenance	4.0
Interceptor and storm sewer improvement costs	7.4
Sediment control and acid mine drainage reduction costs	6.6
Reduction costs for oil spills, water crafts discharge, and miscellaneous	1.0
Added reservoir storage for low flow augmentation	1.0
Total	42.3

In summary, it is in the great American tradition for the American people to work their will through our democratic institutions. Through the processes of public hearings, forums, educational systems, and mass media, the citizen can examine positions and attitudes and make judgments. Decisions should be made by a fully informed citizenry with cost-benefit ratios, alternatives, options, and trade-offs placed before them for consideration. Decisions affecting the quality of individual living should not be made solely by industry, government, eco-freaks, doom mongers, or nationally organized environmental groups. If we reach the point where our policy decisions are made on the basis of only the powerful, organized lobbies of vested interests or we throw up our hands in confusion and lose the battles by default, our nation is in deep trouble.

In the final analysis, the people must be the arbitrators of what constitutes the good life or the decisions on trade-offs and social progress in whatever form. Without an objective EQ Index to assist them in determining with some accuracy the degree to which the environment is deteriorating, the costs, the alternatives, and the trade-offs, they cannot make meaningful judgments.

We can be pessimistic. We prefer to be optimistic. Constructive actions can be taken. Pollution *can* be controlled, strip-mined areas *can* be reclaimed, blighted urban ghettos and wide expanses of highway *can* be made pleasant and productive through proper planning and meaningful, forceful execution of those plans. Clear waters and clean air, green forests and fields, flights of birds, and the sight of wild creatures are among the amenities that make life worth living as compared to a mere existence.

The scientific community has an obligation to assist our decision makers in giving the American people the opportunity to choose between a denaturalized, defiled, and debased existence or a new birth of quality living complete with the preservation of objects of great natural beauty and aesthetic appeal in addition to the restitution and rehabilitation of those renewable natural resources upon which life is dependent. There are those who say science is incapable of providing the technology to clean up our polluted planet. I say the capability is there, but the willingness to devote the time, to pay the price, to establish objective standards of enquiry, and to identify pollution abatement as a high priority domestic and world problem is lacking.

There are those who advocate a no-growth, no further industrial development policy as the only solution to the environmental dilemma. While this may be a worthy objective, such an advocacy is not very realistic. The real solution lies in allocating whatever percentage of our increase in the gross national product is needed to clean up the environment. The construction of a sewage treatment plant or the development of new technology to more economically remove sulfur from coal and oil contributes jobs and additions to the GNP as much as any other productive action.

In conclusion, the question often asked is what can the scientific community contribute to an improved life style through an improved environment? First, the AAAS as the leading representative of the scientific community can help restore the credibility of the scientist and reinstate his position as the fountain of truth by immediately establishing ethics committees for the various resource disciplines. The most outstanding, objective, and respected scientists should be appointed to the committees and specifically assigned the extremely difficult but not impossible task of discerning scientific truths in the arena of conflicting analysis and viewpoint. Those scientists whose position is dictated more by the paycheck than by objectivity or those who make a conclusion and proceed to gather and consider only data that supports that conclusion should be called to an accounting before their peers.

Second, the professional resource manager should immediately embrace an advocacy role in environmental affairs. The scientist should no longer be content with publishing the methods and conclusions of a research effort or handling his working assignment. In this modern day of the activist and public participation in policy determination, the cool, calculated, objective, and expert voice of the true scientist is badly needed in molding public opinion in the proper form and in formulating guidelines that will direct our national policy toward an improved natural environment and a quality life style for all Americans.

REFERENCES

1. The Gallup Organization, Inc., The U.S. Public Considers Its Environment (conducted for the National Wildlife Federation). Princeton, New Jersey, 1969.

2. National Wildlife Federation, National Environmental Quality Index, *National Wildlife,* Aug.–Sept., 1969.
3. National Wildlife Federation, 1970 National Environmental Quality Index, *National Wildlife,* Oct.–Nov., 1970.
4. National Wildlife Federation, 1971 National Environmental Quality Index, *National Wildlife,* Oct.–Nov., 1971.
5. National Wildlife Federation, 1971 EQ Index Reference Guide, Washington, D. C.

USES OF ENVIRONMENTAL INDICES
IN POLICY FORMULATION

Gordon J. F. MacDonald

Member, Council on Environmental Quality
Executive Office of the President
Washington, D. C. 20006

Responsible decision-making in government as in other institutions depends on the availability of reliable information. In areas such as the environment, where emotions can run high, hard facts are often of critical importance. If we are to achieve effective management of our environment, we will need comprehensive data about the status and changes in the air, water, and land. Optimally, these data should be organized in terms of indices that in some fashion aggregate relevant data. At present, our measures are imperfect. The issue of potentially hazardous chemicals, particularly possible substitutes for phosphates in detergents, illustrates both the needs and gaps of information in the formulation of environmental policy.

Last summer, the federal government was faced with making a difficult decision concerning nitrilotriacetic acid — more commonly known as NTA. In such decisions, the Council on Environmental Quality (CEQ) is consulted with respect to the environmental aspects of the problem. As is usual in many of the situations facing the Council, a determination had to be made in a relatively short time as to what sort of policy recommendation would be made on this issue.

A year or so ago when the great wave of adverse publicity about phosphates arose, NTA quickly received public attention as a possible substitute for phosphates in detergents. While possessing many of the beneficial properties of phosphates, NTA would not have the stimulating effect on eutrophication that phosphates often have. Indeed, NTA seemed like such a reasonable replacement for phosphates that at least one company, in good faith, put a sizable investment into facilities to produce NTA on a sufficiently large scale to meet the demands for their detergent products.

This is another example of our society jumping the gun on products that apparently will fulfill a consumer demand, but which have not been sufficiently investigated for their possible side effects on human health and on other areas of our environment. In the fall of 1970, investigations concerning the side effects of NTA were intensified. Just about a year ago, a review to evaluate these effects was conducted by the Public Health Service, the Environmental Protection Agency, the Office of Science and Technology, and our Council. From this evaluation, it became apparent that NTA did indeed have possible adverse side effects that should be carefully investigated before the product received widespread distribution, such as it would if used in detergents. These side effects include a possible chelating action with cadmium and other heavy metals that could result in their abnormal concentrations in human and other biological systems. NTA was also suspected of being carcinogenic.

As a result, the government did not encourage the use of NTA for widespread distribution and pressed for more intensive investigations concerning other substitutes that might replace phosphates in our detergents. The affected industries voluntarily held NTA-based detergents off the market. That was the way the situation stood in August 1971. There was continuing pressure from environmental groups to remove phosphates, even though many detergent manufacturers had already reduced the amount used in their detergents. But no clear-cut substitute for phosphates, other than NTA, was apparent. The evidence on the adverse effects of NTA was not conclusive in all aspects. On the other hand, certain manufacturers urged the approval of NTA, particularly those that had made substantial capital investments to produce it. The federal government had to make a decision on what sort of guidelines relating to this problem should be issued to industry.

It was a tough decision because reliable data were lacking and we had to rely on imperfect assessments. Ideally we should have had the following information:

(1) A definitive model of how eutrophication works in aquatic bodies and the role that phosphates and other nutrients play in the different categories of these bodies, such as lakes, streams, and salt water estuaries, as well as under various conditions of hardness. Information on how to determine the limiting nutrients in each of these bodies would have been particularly useful.

(2) The amount of phosphates from detergent wastes that presently gets into each of these aquatic bodies as compared to other phosphate sources and the amount that planned improvements to our industrial and municipal sewage facilities would remove.

(3) The amount of phosphates that has been eliminated from discharges into our aquatic bodies due to control measures recently instituted and the resulting effect on the eutrophication process in the aquatic bodies. These controls include the voluntary measures exercised by the consumer, the controls promulgated by certain local governments such as Suffolk County in New York, and the phosphate extraction methods implemented in some sewage treatment plants.

(4) The adverse effects of NTA and of NTA compounds on human health and on the other areas of the environment. Included would be the amount which could be safely emitted to the environment, that is, the threshold limit value that must be achieved before we start detecting adverse effects, in the long as well as in the short run.

(5) The same adverse effects information on other possible substitutes for phosphates in detergents and the amounts of these substitutes which would have to be used in the detergents.

(6) The probability for success of technology in removing phosphates, NTA and other substances of possible use in detergents from sewage, in both municipal and industrial processes, and the time required to achieve this.

(7) The costs to the consumer and to the taxpayer for all the above.

A sound research program could produce part of the desired information, such as the success potential of various technological techniques to remove these phosphates and other chemicals in the sewage treatment process. However, much of it, such as the success of the controls instituted over the past two years to remove phosphates from detergents and the resulting effects on the eutrophication process in the receiving bodies of water, would have to be obtained from actual experience and the empirical data obtained from this experience. There has been in fact a substantial experience in various control schemes, but this experience has not been quantified. In the ideal situation, data on water quality would be compiled and analyzed, and the results would be presented to the federal policy formulators in the form of indices relating to water quality. These indices would assist them in their efforts to provide options concerning new policies or changes to existing policies which might be made in order to minimize the impact of man's activities on the eutrophication process.

An important feature of these indices is their assistance in showing the ability of our technology to perform the job listed in each option; the cost, both economic and social, to implement these options; and the lead time necessary to initiate these programs. Also, information on the possible incentives – for example, regulations, tax breaks, and the like – would be extremely helpful.

As you recall, the government did make a policy decision on NTA last September. It was not to foster its use, but rather to allow the use of phosphates in detergents at the lowest levels possible, provided other additives did not pose a health hazard; phosphates at that time appeared to be the "least bad" of the possible detergent additives. Our recommendation had to be made; so, based on the best of the inadequate information and expert opinions available at that time, we made it. This is not the most satisfactory method of operation, but unfortunately with the current inadequacy of data and indices in the environmental field, we must do this more often than we like.

The above example illustrates the usefulness of environmental indices to policy formulators. Other important uses are to assist the informed public in assessing how well the programs dealing with our environmental problems are progressing and to assist acientists and engineers in their development of the technology essential to solving our problems. A prime requirement for indices is that they must be in a form that is easily understood by the different groups using them. The cost of living index used to gage our economy approaches this goal. A second requirement is that the indices must be representative of the geographic area and time period which they are supposed to cover. Here a trade-off must be made between the resources and time required to obtain sufficiently representative samples and the need and timeliness of the resulting information. A third requirement is that indices must be reliable. We are uncomfortable in situations where we are dealing with tenths of a percent to indicate trends, such as for unemployment, and find out that other factors, such as seasonal adjustments, might introduce uncertainties of greater than 0.1% to the apparent values given by the index.

Another requirement is for ease in collecting the data required for the index, evaluating those which are directly representative and those which are not, correlating this data, and processing the useful data into the index. Of course, we would like to use existing information collection and analysis systems to accomplish this to the maximum possible extent.

Section 204 of the National Environmental Policy Act of 1969, which established our Council, describes one of our duties as:

"To gather timely and authoritative information concerning the conditions and trends in the quality of the environment both current and prospective, to analyze and interpret such information for the purpose of determining whether such conditions and trends are interfering, or are likely to interfere, with the achievement of the policy set forth in Title I of this Act, and to compile and submit to the President studies relating to such conditions and trends;"

In March 1970, President Nixon in Executive Order 11514 further defined the duties of the Council in this area to:

"Promote the development and use of indices and monitoring systems (1) to assess environmental conditions and trends, (2) to predict the environmental impact of proposed public and private actions, and (3) to determine the effectiveness of programs of protecting and enhancing environmental quality."

This amounts to quite a challenge when we examine all of the activities of man which have effects on the environment. In many of these areas, there is a great deficiency in the data which we need for determining and understanding these effects. In others, there is an overabundance of data — so much that we do not have the capability to process and interpret it. These data, such as that from some satellites which continuously accumulates on magnetic tapes and cards and fills up vast volumes of valuable storage space, contribute to the growth of a new type of pollution — information pollution. We have got to make sure that the data we collect are data which we can efficiently use and which will not saturate our collection and analysis systems. We must take a hard look at existing systems to insure that they meet this requirement. We must design future collection and analysis systems to achieve this end.

As a first step in building a national environmental monitoring system, the Council commissioned a contractor in the summer of 1970 to develop a system design concept for monitoring the environment of the nation. His final report was rendered in April 1971 and had been commented on by other federal agencies and public groups. The report provided us with certain recommended indices to represent the state of the environment, described the data needed to derive these indices, reviewed current monitoring programs, and considered the alternative methods to obtain and utilize these data. The contractor developed a system to evaluate the indices' environmental impacts, utility, and cost and used it to obtain a relative priority ranking of each index. Although many of the value judgments had to be made arbitrarily, the report performed a real service in giving us something to constructively criticize, whereas before we had nothing.

One of our Council's primary interests is to develop environmental indices that assist in policy formulation and that also inform the general public on the state of the environment and how it has changed over time. We would like to come up with a system similar to that of the Council of Economic Advisors in which indices such as the cost of living index, the unemployment index, and the gross national product provide information on the state of our economy. However, we do not feel that we have sufficient knowledge at the present to define what overall indices we need to represent the total environment. To obtain a better handle on this problem, we are now concentrating on a few selected areas of the environment and are developing detailed indices along with the pertinent reliability factors for these areas.

The six areas which we have selected are air pollution, water pollution, land use, recreation, wildlife, and pesticides. The Council has a contractor in each of these areas who is working with the appropriate federal agencies. Federal agencies in general have most of the available data which might be used as input for indices; however, agencies now generally use this data for different purposes. The available data must be identified, methods of aggregation developed, and procedures established for a continuing effort. These tasks are the responsibilities of the contractors. After the due date of the contractor reports on April 15, we plan to go directly to the pertinent agencies and ask that they improve and, at periodic intervals, update these indices. Also, we hope that these will form the nucleus around which to build the system of indices in other environmental areas so as to eventually have a set of indices that will fulfill the needs of the national policy makers and of the informed general public.

A year ago, a new agency – the Environmental Protection Agency (EPA) – came into being. EPA is the pollution control policeman of the federal government.

EPA fulfills the role of establishing criteria and necessary standards pertaining to environmental issues and of seeing that the measures to implement these standards are promulgated and enforced. In conjunction with the other federal agencies, EPA now provides us with the constructive antagonist– protagonist ingredient in environmental matters. Historically, this ingredient has played a key role in so successfully shaping our country's policies in other vital areas.

With EPA on the scene, it is quite logical that it should play a major role in establishing the environmental monitoring and information system. It assumed control of some good monitoring systems as a result of the agencies which came under its control in the federal reorganization which created EPA. These include the National Air Sampling Network from the old National Air Pollution Control Administration and the Pasteurized Milk Network, to monitor several radionuclides, from the old Bureau of Radiological Health of the Department of Health, Education, and Welfare.

EPA is now in the process of integrating the monitoring systems that it inherited, along with the information output from the systems of other agencies, into a form best suited to its mission. It could presently use the data from environmental monitoring systems to accomplish the following:

(1) To help enforce environmental quality standards for air, water, radiation, pesticides, and solid wastes.

(2) To develop new standards and refine existing standards.

(3) To analyze and report the early buildup of hazardous pollutants.

(4) To ascertain whether segments of the environment are improving or declining in quality from one time frame to the next.

(5) To assist the Executive Branch in establishing priorities for environmental work and in deriving the cost and benefits of alternative EPA actions.

I think we all realize that we cannot separate environmental considerations from the activities of all agencies in the government and create one super agency to deal with just this task alone. If we did, there would be very little work left for the other agencies since environmental considerations enter into just about everything we do. So, through the environmental impact statement mechanism which was established by the National Environmental Policy Act, all agencies are required to consider environmental factors as an integral part of their actions. Until two years ago, the main considerations in agency actions had been the mission, available technology, and the cost. Now a fourth major consideration has been added — the environment.

As I mentioned earlier, the federal government has a number of existing agencies that are working on environmental problems and that have developed monitoring and information systems applicable to this work. An example is the surface water monitoring network developed by the U.S. Geological Survey in the Department of the Interior. We do not have the resources to build a new environmental monitoring and information system from scratch. We must make use of the systems which have been developed by other agencies of the federal government, like the USGS network. Of course, most of these systems were developed to focus on just one medium or pollutant, whereas the effects of pollutants on man and on the rest of the environment come from a combination of pollutants traveling through several media. Quite often synergistic effects and processes create byproducts that are even more harmful than the original substance itself. An example of the latter is methylmercury, which many think is produced in marine sediments from man's disposal of metallic mercury. This methylmercury has a much higher absorption factor in the human body than does the metallic mercury and hence is much more dangerous to human health when it enters the human food chain from these aquatic areas.

We must insure that the existing federal assets are used more efficiently. To help achieve this, the Office of Science and Technology sponsored a study on Environmental Quality Information and Planning Systems, commonly known as SEQUIP, in the year preceding May 1971. This study has fairly well identified what the present assets of the federal government are in the information field relating to the environment. This is indeed a valuable contribution to the present CEQ and EPA efforts.

To summarize my remarks, we need better and more reliable data to understand the effects of man's activities on the environment and to determine what possible things we might do to ameliorate the adverse effects. This data must be presented in abbreviated but meaningful form to policy and decision makers, to the informed general public, and to scientists and engineers so that all three groups may understand them and make use of them in fulfilling their respective roles in preserving and enhancing the quality of our environment. Our Council makes policy recommendations to decision makers and coordinates efforts in the field of the environment among federal agencies. We need indices covering the whole spectrum of environmental activities as soon as possible so as to make our recommendations more meaningful and more reliable. These indices will assist us to define the various options involved in policy decisions, to include their economic and social cost and their implications on other activities, and to assess the success of present policies in dealing with some of these problems.

URBAN-ENVIRONMENTAL INDICATORS IN MUNICIPAL AND NEIGHBORHOOD POLICY PLANNING AND DECISION MAKING

John Berenyi

Office of the Mayor
City Hall
New York, New York 10007

During the past few years, national attention has focused for the first time on the deteriorating condition of urban environment. It has become public knowledge that environmental decay not only degrades the quality of life, but more importantly threatens health and property as well.

Millions of Americans living in large cities, medium-size towns, or the suburbs surrounding our large urban centers have increasingly asked the question: What has our government been doing to clean up the environmental mess? The governments they refer to most often are not located in Washington, Albany, or Sacramento but in New York, Philadelphia, San Francisco, Atlanta, and Pittsburgh. And the environmental mess they cry about includes sanitation, air and water quality, lead poisoning, noise, and traffic congestion.

If we look at American and world cities in the context of national and international society, we can conclude that we have not done well in protecting the environment. New York and other cities have a long history of economic growth at the expense of environmental quality.

The streets of many of our cities have a long established and often justified reputation for filth. The quality of the air contains excess amounts of sulfur oxides, carbon monoxide, and other pollutants. Traffic congestion resulting from the increased number of automobiles occupying city streets creates daily nightmares for the rush-hour traveler and contributes to the expansion of noise in our atmosphere.

Municipalities, the governmental structures closest to the people, had to respond to these vast environmental problems choking the citizens. Through legislative change, organizational restructuring, and relevant city-wide programs and projects, New York City has been in the forefront, with our limited resources, in the national fight to clean up the environment.

In a 1966 reorganization of New York City government, the Departments of Air and Water Resources, as well as the Sanitation Department, were brought together under the umbrella of the Environmental Protection Administration that later served as a model for the federal EPA and governmental reorganization in other cities and states.

After revising some of the environmental quality standards — in numerous areas, our laws are tougher than applicable state and federal standards — the City began to develop environmental, transportation, and health programs to reduce the level of pollution in the environment.

On all levels of decision making, whether in City Hall, the Bureau of the Budget, the Department of Air Resources, or at neighborhood centers, we rely on and use appropriate environmental indicators and indices, for they are important inputs in the policy analysis and evaluation phase.

Table 1. Summary of routine air analysis

Material analyzed	Method used	Instrument used	Details of reporting
Dustfall	Gravimetric	Stanchion analytic balance	Tabulation
Smokeshade	Transmittance	RAC; tape sampler; spot evaluator	Two-hour values; tabulated monthly; daily graphs.
Oxidants	Saltzman sulfamic acid (chromium trioxide)	Auto bubbler colorimeter	Hourly values; tabulated monthly.
Nitrogen dioxide	Diazotization coupling (Jacobs Hochheiser)	Auto bubbler colorimeter	Daily report; tabulated monthly.
Nitric oxide	Diazotization coupling (permanganate)	Auto bubbler colorimeter	Daily report; tabulated monthly.
Oxidants	Sulfamic acid	Auto bubbler colorimeter	Daily report; tabulated monthly.
Hydrogen sulfide	Methylene blue	Auto bubbler colorimeter	Daily report; tabulated monthly.
Sulfur dioxide	Peroxide titration	Auto impinger burette	Daily report; tabulated monthly.
Carbon monoxide	Molybdenum blue (NBS)	MSA tube	Daily report; tabulated monthly.
Aldehydes	Goldman–Yagoda iodine titration	Burette Greenburger–Smith impinger	Daily report; tabulated monthly.
Dust count	Micro analysis membrane filter	Microprojection micro analysis holder	Daily report; tabulated monthly.
Carbon monoxide	Infrared	Mine Safety Appl. recorder	One-hour values; tabulated monthly; daily graphs.

As an example of how this process works, a typical policy development will be traced through.

New York City conducts routine air sampling analysis on such materials in the air as nitric oxide, sulfur oxide, aldehydes, carbon monoxide, dust, and many others. (See Tables 1 and 2 for complete list.) This analysis is conducted at 38 stations that are part of the aerometric network of New York City. The indicators and indices in use provide data in various time periods from all parts of the city, as shown in Fig. 1 and Table 3. The functions of the system are:

1) To objectively measure progress toward air quality goals

2) To indicate where further abatement steps are required

3) To define air quality in all parts of the city

4) To provide information which could activate the Alert-Warning System

5) To provide a daily report to the public

6) To provide a data bank for epidemiological and other researchers

7) To provide data for constructing mathematical models

Table 2. Summary of Routine Air Analysis

Material analyzed	Frequency	Types of sampling (flow rate)
Dustfall	Calendar month	Continuous
Smokeshade	2–hour period; 370 per month	Sequential (0.25 ft^3/min)
Sulfur dioxide	Continuous	Continuous (3.0 liter/min)
Sulfur dioxide	1–hour period; 24 per day; 5 days/week.	Sequential (0.9 ft^3/min)
Carbon monoxide	Continuous	Continuous (4 liter/min)
Particulate matter	24–hour period; 5 days/week	Grab (40 ft^3/min)
Nitrogen dioxide	1–hour period; 24 per day; 5 days/week.	Sequential (3 liter/min)
Oxidants	1–hour period; 24 per day; 4 days/week.	Sequential (3.0 liter/min)
Oxidants	1–hour sampling; 2 times/day, 5 days/week.	Grab (3.0 liter/min)
Hydrogen sulfide	2–hours/day; 5 days/week	Grab (3.0 liter/min)
Sulfur dioxide	1/2–hour/day; 5 days/week	Grab (1.0 ft^3/min)
Carbon monoxide	2 per day; 5 days/week	Grab (100 cc/min)
Aldehydes	1/2 hour/day; 5 days/week	Grab (1.0 ft^3/min)
Dust count	7 min/day; 5 days/week	Grab (10.0 liter/min)

The statistical data generated by the aerometric network are summarized and evaluated by the Division of Technical Services of the Department of Air Resources. The results provide the objective analyst with information as to how New York City and its neighborhoods meet federal, state, and local standards.

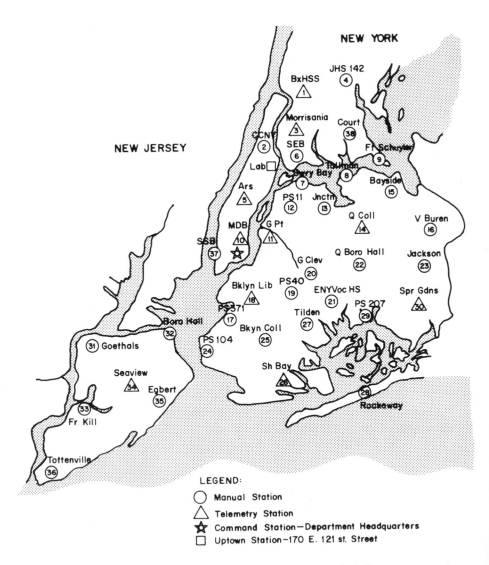

Fig. 1. Aerometric network of New York Department of Air Resources.

Table 3. New York City Aerometric Network

Sector number	Type of station[a]	Site and location	Borough	Station elevation (ft) above street	above sea level
1	T	Bronx High School of Science 205 St. & Goulden Ave.	Bronx	50	200
2	M(CN)	City College, City U. of N.Y., Convent Ave. & 141 St. (Tech. Bldg.)	Manhattan	75	175
3	TA(CN)	Morrisania Health Center, 1309 Fulton Avenue	Bronx	65	135
4	M(CN)	Junior High School No. 142, Baychester & Schlieffelin Avenues	Bronx	45	125
5	TA(CN)	Central Park Arsenal Bldg. (Dept. of Parks), 5th Ave. & 64th Street	Manhattan	45	95
6	M	Samuel Gompers High School, 455 Southern Blvd.	Bronx	65	95
7	M(CN)	Bowery Bay Pollution Control Plant, 41 St. & Berrian Blvd.	Queens	60	75
8	M(CN)	Tallman Island Water Treatment Plant, 127th Street & East River	Queens	40	55
9	M(CN)	Fort Schuyler, USN, MC, RTC	Bronx	30	40
10	T(CN)	Mabel Dean Bacon High School Annex, 240 Second Ave.	Manhattan	100	125
11	T(CN)	Greenpoint Pollution Control Project, 301 Greenpoint Ave.	Brooklyn	25	35
12	M	Public School No. 11, 56 St. & Skillman Avenue	Queens	70	165
13	M	Dept. of Health, 34–33 Junction Blvd.	Queens	50	125
14	TA	Queens College, City U. of N.Y., 65–20 Kissena Boulevard	Queens	45	145
15	M	Bayside High School, 32 Ave. & Cpl. Kennedy Blvd.	Queens	60	155
16	M(CN)	Van Buren High School, Hillside Avenue & 231 Street	Queens	50	150
17	M	Public School No. 371, Fourth Ave. between 36 St. & 37 St.	Brooklyn	45	80

Table 3 (continued)

Sector number	Type of station[a]	Site and location	Borough	Station elevation (ft)	
				above street	above sea level
18	TA	Brooklyn Public Library Grand Army Plaza & Eastern Parkway	Brooklyn	100	235
19	M(CN)	Public School No. 40, Ralph & Marion Avenue	Brooklyn	50	100
20	M	Grover Cleveland High School Himrod St. & Grandview	Queens	60	170
21	M	East New York Vocational High School, 1 Wells Street	Brooklyn	60	95
22	M	Queens Borough Hall 120–55 Queens Blvd.	Queens	40	120
23	M	Andrew Jackson High School Francis Lewis Blvd.	Queens	60	110
24	M	Public School No. 104 9115 5th Avenue	Brooklyn	55	85
25	M	Brooklyn College Bedford Ave. & Campus Rd.	Brooklyn	45	75
26	T	Sheepshead Bay High School 3000 Avenue X	Brooklyn	40	45
27	M	Tilden High School, Tilden Ave. between 57 St. & 59 St.	Brooklyn	60	85
28	M	Rockaway Pollution Control Project, Beach 106 St. & Beach Channel Drive	Queens	34	35
29	M	Public School No. 207, 88 Street &159 Avenue	Queens	45	55
30	T(CN)	Springfield Gardens High School Springfield Blvd. & 144 Ave.	Queens	40	65
31	M	Goethals Bridge Plaza	Staten Island	40	50
32	M	Borough Hall–St. George	Staten Island	50	110
33	M	Fresh Kills Sanitary Landfill, Muldoon Avenue	Staten Island	35	45
34	TA(CN)	Seaview Hospital 460 Brielle Avenue	Staten Island	60	360
35	M	Egbert Junior High School, 2 Midland Avenue & Boundary Lane	Staten Island	45	55
36	M	Tottenville High School, Academy St. between Amboy & Hylan	Staten Island	55	130

Table 3 (continued)

Sector number	Type of station[a]	Site and location	Borough	Station elevation (ft) above street	Station elevation (ft) above sea level
37	M	SS *John Brown*, Pier 42, Morton St. & Hudson River	Manhattan		
38	M	Criminal Court, Part 5. 1400 Williamsbridge Rd.	Bronx	45	90
00	M	Laboratory 170 East 121 Street	Manhattan	15	35

[a]TA = Telemetering station in Alert-Warning System
T = Telemetering station
M = Manual station
CN = Corrosion network station

This information, standards vs. compliance, is passed with other information on the programs of the department to the New York City Environmental Protection Administration. The function of this agency at this point is to provide program alternatives and options for further improving environmental quality which could be evaluated by the Bureau of the Budget and the Mayor's Office within the fiscal and political constraints facing the Mayor. At this juncture, environmental quality indicators become an input in the political decision-making process.

Recent plans for a Madison Avenue Mall have the broad objectives of recapturing space in central Manhattan for outdoor life, lessening air pollution, and unsnarling the city's traffic jams. In the public discussions and debates that follow the presentation of such plans, Mayor Lindsay and his associates need and use all the available data to convince business groups, community organizations, and civic clubs that this plan will indeed improve the quality of life of the city without causing serious hardship to the various special interest groups. The tools and techniques that scientists, engineers, planners, lawyers, and others have developed become the vocabulary for communicating the needs of the citizens to municipal decision makers, and these are used by policy makers to make a case for an important program or project.

Progress on the environmental battlefield is difficult to achieve, but our city is making some solid strides in the right direction. Because of tough air quality standards and enforcement procedures, since 1966, for example, there has been a 30 % reduction in particulate emissions and a 60 % reduction in sulfur oxide emissions. The number of acceptable days in 1970 was 24 % compared with 10 % in 1969. And the number of unhealthy days dropped nine percentage points in 1970 from the previous year.

There are many other indicators that the city uses and new ones it needs for its environmental and municipal policy planning and monitoring process. Some of the frequently used indices include:

1) Bacteria counts in water

2) Frequency of collection by the Sanitation Department

3) Tons of garbage handled by a sanitation worker

4) Number of abandoned cars per police precinct

5) Significant level of lead in the blood

An interesting environmental indicator that tells us something about the socio-psychological stability in a neighborhood is the number of illegal fire hydrant openings and the corresponding water pressure in the area. It is not surprising that the number of illegal hydrant openings in the period of July 19-25, 1971 was 147 in the north part of Manhattan, which includes Harlem, compared with one illegal hydrant opening in the south part of Manhattan where the elegant Sutton Place, parts of Fifth Avenue, and the fashionable Gramercy Park sections are located.

Finally, the signs are that the future demands for urban—environmental indices will come from cities and their surrounding neighborhoods. The indicators that could project and predict second-order consequences of governmental action, as well as allow the decision maker to judge the probable costs and benefits of policy change, will be greatly sought after. Government, media, and the general public will increasingly accept those indicators that communicate some information about relative progress and achievement within the context of social change. Measuring performance as well as productivity of government against established goals and objectives will tell us a great deal about where we are in relation to where we have been and allow us to plan for the direction we should be going to in the future.

ACKNOWLEDGMENT

Eileen Brettler and Cheryl Garnant performed valuable research for this paper.

BIBLIOGRAPHY

Berenyi, J. (ed.). *The Quality of Life in Urban America—New York City: A Regional and National Comparative Analysis,* Vols. 1 and 2, Office of the Mayor, New York, 1971.

THE UNCOMMUNICATIVE SCIENTIST:
THE OBLIGATION OF SCIENTISTS TO EXPLAIN
ENVIRONMENT TO THE PUBLIC

Peter Hackes

Correspondent, National Broadcasting Company
4001 Nebraska Avenue, NW
Washington, D.C. 20016

I shudder to mount this podium in the presence of so many science luminaries. Your credentials are formidable: science degrees, published works, and many years of research and teaching in one or more of the scientific disciplines.

My own achievements in science, I'm afraid, are limited to some mediocre grades in college physics and biology, plus a pat on the back from a professor when I finally learned to use a parallel rule in an engineering drawing course.

Later, as a graduate student at the University of Iowa, I was sent out – as a reporter for the Journalism School publications – to "investigate" the Physics Department. Seems there was a gentleman there whose pride and joy at the time was a contraption called a Van de Graaff generator. No one in the journalism department knew what a Van de Graaff generator was. My editor, I'm sure, thought it was probably a machine that made Van de Graaffs – whatever they were.

So I sought out this science creature ... who turned out to be a most enjoyable individual, who when he wasn't busy generating Van de Graaffs took time to explain to me (in English, not algebra) what he was doing. His name turned out to be Van Allen, which at the time meant very little to me.

I did a short piece on Dr. Van Allen and his generator. I must confess I was a much wiser reporter for it. He was (and is) the kind of scientist who combines an extremely human quality along with his towering scientific bent. He exhibited a real desire to answer my multitudinous questions in a way that I could relate to what he was doing. Not all scientists have the ability or even the willingness to put their work into words of two syllables so an unversed reporter can understand. Not all scientists have the ability *or* the desire to get across what they are doing to the general public ... whose representative I (as a reporter) feel I am.

True, science writers should have at least a modicum of science in their background, but the plain fact of the matter is that a good many of us don't. I'm sure this can be the fault of the reporter himself. It can be the fault of the school or schools he attended. It might even be the fault of his journalistic organization which probably has neither the funds nor the inclination to send him on a leave of absence to at least dabble in a couple of science courses.

The situation shows improvement now and then when a person who has had some years as a scientist decides to write for the general public as a reporter for a science publication or becomes a syndicated columnist. There are those of that ilk who in recent years have turned themselves over to radio and television ... to become the Howard Cosells of such things as total eclipses and photos from Mars ... much to the horror, I'm sure, of their science colleagues.

But these gentlemen are rare and valuable. The stark reality is that – with some few exceptions – the reporting of scientific progress is (of necessity) left to reporters who wish they knew a great deal *more* about the fields they cover. It's a fact of journalistic life, but one which need not be self-defeating. Reporters, after all, cover fires, and they're not firemen; they cover Senate sessions and they're not senators; they cover wars and they're not soldiers; they cover the United Nations and they're not professional diplomats. So why can't they be expected to cover the science field even if they're not scientists?

The answer is: they can. But they can't do it all by themselves. Particularly in my area – radio and television – it's hard for an outsider to grasp what's going on. It's even harder to translate it into words – and *pictures* – that can be understood by the public. But that's my job.

Aside from an occasional news conference held by a science organization in Washington, my contact with the world of science was almost nil until the summer of 1957 – the start of the International Geophysical Year. My editor decided it needed covering and sent me. It was an eye-opener. I was thrown into science meetings at which I understood hardly a word, but I had to sink or swim and before long I had pulled myself up to where my nose was just enough above water so I had a rough picture of what it was all about. Then and only then did I feel well enough versed to pass along some of what I had been seeing and hearing to those who were watching and listening.

Shortly thereafter, I was assigned to the space beat by my network. My IGY coverage had been good groundwork. Unlike some of my reportorial colleagues, I found I knew where space was! And I had had the exhilarating experience of following the Vanguard satellite project through its many ups and downs. The space program was another challenge. Many an hour, day, and week was spent at what was then Cape Canaveral, at M.I.T., at Houston, and at the various locations where the space hardware was being manufactured.

The environment, when it became an IN thing a couple of years back, was a new challenge. Back to the books again. Back to old science friends and new ones – to engineers who were expert in this or that – all of it aimed at trying to elevate my meager knowledge in this field.

I found, for example, that environmental indices do not lend themselves to instant television depiction. Reporting things that often cannot be seen sometimes becomes a problem. But we *can* illustrate the *effects* of bad environment, in the hope that such reportage will lead to something better.

Our group leader for this discussion of environmental indices wrote me some months ago outlining the purpose of this gathering and saying (among other things) that there were two questions that kept nagging him: First, asked William Thomas, "Are we concerning ourselves with esoteric issues that do not really concern the public?" And second, Tom asked, "Would our task be simpler if we had constant feedback from persons not actively involved in science?" "Science," he pointed out, "now has a badly tarnished image — an image created primarily by scientists' lack of communication with the public."

In response I would say these issues — especially the matter of developing environmental indices — *vitally* concern the general public, whether or not they are currently aware of the dangers of *not* being concerned. Second, it's my job, and the collective task of many others in my field — our *responsibility* — to pass along to the public whatever information we can glean from you in the science community regarding the environment and what effect it can have in the day-to-day life of the community at large and also the long-range outlook. But before we can "smarten up" the general public, there are a lot of us reporters who need "smartening."

I must confess that when I first became interested in reporting environmental matters I approached the field just as a reporter would approach any other assignment: get the facts and write the story of what's happening. As simple as that. That's the way we do it at City Hall. That's the way we cover a fire or an accident. Those are basically the procedures for reporting *any* kind of story.

But it wasn't long before most of us realized that reporting on ecology wasn't the same as the everyday murder where the crime is uncomplicated, the victim is easily defined, the motives are clear, the perpetrator is a blackguard, and the jury's decision can go only one way.

Many of us started off in that direction. All went well at first. It didn't take us long to find the pollution: the smog was unbelievable; you could almost walk on many rivers; city noise actually was beginning to make people deaf; and localities were beginning to find that not only were their community dumps contaminating the soil and water in many areas but the communities were beginning to run out of acceptable dump space.

These were the simple, easy-to-report stories that occupied many of us in the early days of environmental awareness. There was little doubt about what was going on and how to clean it up. Cause — effect — and remedy. Then not long after it all began, what had appeared to us to be an open-and-shut case started taking on something of a different shape. Phosphates, for example. We all knew, because eminent scientists had told us, that phosphates in laundry detergents were the prime culprits in killing off some of our rivers and streams. Phosphate nutrients from waste water had fostered algal growth, which had used up oxygen in the water, which in turn had killed off thousands and thousands of fish and other aquatic life.

To correct the situation, a fairly straightforward step was needed: get rid of the phosphates in detergents. Here again we turned to scientists, the people who had pointed out the problem in the first place. In a matter of months — or so it seemed — they had presented us with a substitute product whose non-scientific abbreviation was NTA. Fine. The detergent companies went wild over NTA. It looked like the answer.

But again it was a scientist who blew the whistle on NTA, which he found may have been producing cancer and creating birth defects among laboratory rats. He admitted that the experimentation was incomplete and its conclusions a bit iffy. Yet, as charged by the bill which created it, the Environmental Protection Agency did what it had to do in such an emergency situation; it got the industry to agree to a voluntary ban on NTA.

So it was back to the drawing board for the detergent scientists. Instead of NTA, they'd add more caustics to their product ... anything, really, to clean up those dirty collar rings and make those blue shirts come out whiter than white. Another few months of happiness in the detergent industry, until scientists found that substitutes for NTA might produce problems as great: the caustics were giving people — especially babies — skin rashes. And when the stuff was taken by mouth — as inquisitive children may do — it produced a most unhappy effect on the windpipe and anything else it touched going down.

By this time, at least one laboratory that had been researching NTA substitutes for some time had reported they had made no progress toward a really good substitute.

So ... back to science again. This time we witnessed the interesting spectacle of two leading government figures at a news conference telling us (and through us telling the nation's housewives) that detergents really are all right to use — phosphates and all, that the important thing to do is to have each community build a tertiary sewage treatment plant to take care of phosphates as they are washed down the line. (The fact that the construction of sewage plants takes years and years and billions and billions of dollars must have escaped the two men momentarily.) We were told it would be best for detergent makers to put out at least two kinds of washday miracles: medium phosphate content for hard water areas and low phosphate content for soft water areas.

But even before this final nostrum was passed along, many communities had begun work on anti-detergent or anti-phosphate laws, many of which are now on the books despite some ardent lobbying by the detergent-makers who, of course, are now supported by the government's pronouncement (in effect) that phosphate detergents aren't really good for the environment ... but they're not really bad for us ... so just be careful what you buy.

Off in the wilderness, it seemed, one could barely hear the murmur of still another group of scientists repeating what they had been saying for lo these many years: why not try soap?

The phosphate story is a good one to point up the problems we all face, especially those of us whose job it is to try to pass along to our readers and listeners the very best information we can find on the environment. What we needed was some good, solid advice — not alone from the government, not from detergent hucksters, and not even from doomsday environmentalists, but from some science types with no particular axe to grind. It was hard to come by.

Nor is the detergent fiasco the only environmental area in which we who want to communicate such things have run into real problems — not so much in informing the general public but in finding out *what* we should be communicating. Let me run down a few others:

Remember "Get the lead out!"? It was a campaign begun by some scientists, a campaign soon picked up by environmentalists and politicians, with government encouragement even to the point of proposing a tax on leaded gasoline. The proposal was to phase out lead in gas because that's what was causing smog. Along came another group whose experiments indicated that eliminating lead from gasoline might even create more smog in some areas — not less.

Remember the outcry against mercury? It put the swordfishermen out of business. For a while, it stopped the tuna boats. That too was a science-generated government ban. Based on more information, it wasn't long before the government relaxed its ban against tuna fish. Comes now a scientist who tells us the amount of mercury in our environment is not on the increase as we had thought but is, in fact, decreasing. He goes so far as to theorize that the presence of traces of mercury may even be essential to life, although he admits "the whole field is in ignorance."

I think you can get a feel for the problem that faces us, the people who must go to science for the answers. What *are* the answers? Is there only one? If there are three answers, which is most correct? For how long is today's environmental edict likely to remain unchallenged before we move along to yet another?

Were the scientists who opposed the Amchitka blast (on environmental bases) *accurate* in their predictions? Can the scientists who successfully defeated the SST (at least in part on environmental grounds) be any more accurate in their predictions of ecological catastrophe than those who took the other side, saying nothing bad would come from fleets of SST's? Is the fast-breeder nuclear power reactor an environmental *menace* or a gadget that will save us from extinction?

One possible substitute for the present-day gasoline engine is a motor that runs on natural gas. (EPA Administrator William Ruckelshaus has a big tank of cooking gas in the trunk of his black government car.) But how can scientists tell us to use gas instead of gasoline when they (of all people) should know how short we are of natural gas supplies? (It's been estimated that to run only half the nation's autos would take *all* the natural gas we have.)

Where are the scientists to give us an unbiased opinion on whether 1985 is too soon to expect the nation's waters to be clean enough to swim in (Senator Muskie notwithstanding)?

Where are the scientists to tell us — in an unbiased fashion — whether the auto makers are correct in saying it's impossible, technologically, to clean up auto exhausts by 1976 — the date by which they'll all be breaking the law if they don't?

Where are the scientists who'll give us an unbiased view of the nuclear power picture? Are we, in fact, risking environmental ruin if we allow heated water to return to rivers and streams? Or is it true that the scientific approach to this problem shows that thermal pollution has been played up out of proportion and that, in fact, for some species of aquatic life, heated water has proved a boon to increased reproduction and enlarged size — just what the local fisherman has been looking for!

I don't mention these areas of apparent scientific difference of opinion to put down men of science. Just the opposite. The word of the scientist, in environmental matters as with nearly everything else in our daily lives, is essential to those of us groping our way through the maze of ecological pros and cons. There are no obvious answers to the questions I've put. In most of these areas there is more than one answer.

Science is needed to guide us so that we in turn can guide the public. Not that the farmer, the bricklayer, the man who runs the local grocery store, or the grammar school teacher are stupid about things environmental. Nor are they without a good deal of basic knowledge about what makes the whole scheme of things tick right along. There's a great deal more sophistication among the American people these days than they are often given credit for.

But there is ignorance. Great ignorance. Not only among that vast audience we call the "general public" but also, if you will, among reporters. We *need* scientists to give us guidance. We *need* the one-time ivory tower people to come down and tell us what they've been doing. The day of the isolated experimenter who goes off in his little corner to contemplate one tiny segment of life all by himself is at an end. These are days, it seems to me, when the science community has a moral *duty* to itself, the public at large, and to future generations to come forward with every scrap of knowledge that might be useful. Nowhere is this any more needed than in the environmental area.

But how to do it? How do scientists (even those who really *want* to) communicate with those whose background is so completely different as to give the impression — one to the other — that each is from a different planet?

I must confess I haven't programmed this matter through my handy-dandy computer. Nor have I run it up the flagpole of scientific research. So my thoughts will have to be those of one communicator whose limited experience has at least presented a few of the problem areas and whose meager efforts to translate science into English have had at least some small success.

Scientists are not newsmen any more than newsmen are scientists. Which is as it should be. Each of us has his problems: the scientist in learning how to communicate; the reporter learning from the scientist *what* to communicate.

Science, after all, is really a gathering together of experimentally proved facts to which is added an analytical thought process. In essence, science is the pursuit of ideas. Have you ever tried to illustrate a thought process for television? This is *our* basic problem: how to take those thoughts out of the head of a scientist and pass them along (with pictures) to a public, whose brain power is there, but whose background may or may not provide the proper basis for understanding.

Until recently, there wasn't much incentive. Scientists appeared to be vying with each other to see how mystically silent they could remain. For whatever reasons of jealousy, empire-building, worry about being misquoted, or fear that other scientists would look askance at someone foolish enough to subject himself to an interview, men of science were more than reticent. They were downright unavailable.

Changing times have changed those attitudes a bit as has, especially, the need to share the government's financial pie, which after all can be divided only into a certain finite number of pieces. At the same time we reporters issued a plea for help. In many instances, these two needs are producing a flow of science news that has never before been apparent. Scientists have realized that they need public support to persuade Congress, for example, that certain projects are worth pursuing. They know that it's the members of the House and Senate who can turn on or *off* the biggest science-oriented money spigot in history.

The times, I feel, are also changing the typical public image of the scientist. He's now a member of the community — rakes his leaves, sends his kids to neighborhood schools, complains about taxes, and watches pro football games (scientifically). More and more often, the modern scientist is taking part in the local PTA, fund drives, and little league baseball. In short, he's becoming a human being.

In this humanizing process, the scientist is also beginning to think thoughts other than those that for so many centuries have chained him to the laboratory table. He's beginning to have thoughts about *why* moves in certain scientific directions are either good or bad. And he's beginning to express those thoughts, where up until recently the scientist was expected to stay out of public view.

This, I feel, is a step in the right direction. I'm one who thinks members of the science community should be encouraged to speak out — not only on *what* they are doing but on *why* they are doing it.

More scientists should be heard, for example, on the need for an anti-ballistic missile, or if they feel otherwise, the *lack* of need for an ABM. Likewise, they should speak out — as some have — on such things as the now-safe possibility of seeking an international treaty banning *all* nuclear tests, because we now have the scientific tools to detect Soviet violations without on-site inspections. (Those scientists who take the *opposite* position should also be heard, of course.) These happen to be two items on which the general public is all but completely unknowledgeable.

In the environmental field, conflicting science views should be heard by the general public, just as in other science areas where there is no unanimity. It should be the moral responsibility of the scientist who lives in a community whose air and water are being fouled by a local industrial plant to speak out *forcefully*. As a living, breathing member of the community, shouldn't it be his *duty* to concern himself with what's going on environmentally and then express himself — if not publicly, at least to the proper industrial or city official?

The magazine *Environment* published a letter recently from a retired science professor from a college in northern New York state. He said that his community has a new secondary sewage treatment plant that costs the taxpayers $184,000 a year to operate, including a $10,000 yearly electric bill. He wonders in the letter: "Since the production of electricity is itself a major polluter, how much *new* pollution is created in trying to clean up an *old* source of pollution?" He concludes by saying that an investigation of the real cost of environmental cleanup might show that in some cases we are merely exchanging one form of pollution for another, at tremendous expense.

We need this kind of thinking *expressed*. Unpopular views should be heard. From time to time, men of science, for example, have told us that a total ban on offshore oil drilling (for many reasons) might, in fact, *increase* the possibilitites of oil pollution along our coasts. Shouldn't that view be presented fully?

Another issue of *Environment* ran an editorial which stated (in part): "There are technologies which should be developed in the public interest but which are not now being developed for reasons of lack of economic attractiveness or economic uncertainties for the potential manufacturers. Examples are low-pollution, low-power automobiles with much-increased crash-resistance and alternatives to the internal combustion engine in autos, buses, boats, and farm appliances." Shouldn't there be a dialogue on these subjects — pro and con — by scientists and engineers who are willing to stand up and tell us what they know? How else can the public tell its elected representatives what decisions ought to be made?

The same right-to-know holds in the area of our discussion today — the field of environmental indices. The lady with emphysema, the child with a croupy cough, and the eldery man with a heart problem *need* to know when an air crisis is brewing. How else, except by being told by scientists, can they learn when there is an emergency and what they should do? Aside from emergencies, just knowledge of environmental problems is the right of the public which has to live with them.

We reporters need the educational under-pinning to recognize routine matters as well as crises. When an environmental emergency arises such as an air crisis over Birmingham, Alabama, the public should have a prior awareness of why the emergency exists or they cannot be expected to respond. The public can and does respond, incidentally, to a non-crisis situation such as voting for sewage plant bond issues and pushing for cleanup legislation, if care has been taken in explaining the situation.

There must be a set of understandable environmental indices to keep people informed of the quality of their surroundings, so that they can see for themselves whether it's getting better or worse — and *why*.

A well-informed public *can* and *will* back up the scientist willing to stick his neck out, *if* it is provided with enough background information to make a point logical through a frame of reference. Surely the person who sits at home and watches the aspirin tablet dissolve inside the glass stomach of a glass dummy on his television screen (and then goes out and buys that aspirin) will be willing to listen to a *real* scientist talking about *real* scientific matters, *if* he can present his ideas in such a simple down-to-earth fashion.

The aspirin commercial points out another truism in the real world of communications in the 1970's: the average television watcher doesn't really need — or even want — to know *why* the aspirin works to relieve his headache. It's enough to show him that it does. By the same token, he doesn't have to know exactly how a smokestack scrubber works or what an auto exhaust pipe cleanup mechanism looks like or what a nuclear power plant cooling tower consists of. He can be told, logically, how they operate and can be advised, logically, how necessary they are, without being given dull exhaustive detail. The same, of course, is true in other branches of science: the average citizen doesn't have to know how a laser beam operates to understand some of the things it can do for him. He doesn't have to have a working knowledge of holography to be shown how it can be used to help him as a tool for improving his life.

That's where people in my profession can be helpful. We're no smarter, scientifically, than the average person whose major occupation is reading the daily comics. But we do have one small knack (or we should): the art of being a surrogate for that person by asking questions of experts which *he* would want asked, so the information is understandable to him.

No matter what the story — even hard-to-illustrate ecological items — there are ways of telling it without compromising the scientific insistence on accuracy. There are ways of relating the basics of an on-going experiment without giving away detailed information to scientific rivals, which could be disastrous to a hard-working team of experimenters.

Most reporters realize that research is fundamental to scientific progress. We know that publicity about research often has to remain a bit vague until you scientists begin to see the light at the end of the test tube. We don't mean to probe. We don't mean to pry out any final answers before they become finalized.

But I think it's only fair (expecially when there may be a great deal of public tax money involved) to wheedle out a progress report now and then. The general public wants science to let it in on what's going on, why an experiment is being carried on, what has motivated the scientist to conduct it, and in some general terms what he sees as the possible future outcome of his work and how it might relate ethically to the rest of the world community.

We're not spying. We realize the sensitivities involved. We understand that a person's scientific reputation can be wrapped up in a series of experiments. But where those experiments involve the general public, it now behooves the scientist to make sure he's communicating. If he doesn't, he merely perpetuates his unpleasant public image.

I am well aware that not every reporter is science-oriented. I'm aware that some of us don't have the background to grasp the import of what the scientist is saying. Here, I feel, it's the duty of both sides to try a little harder. Only in this way can we avoid some of the mistakes we've all seen – the reportorial mistakes which are the bane of our existence.

Science writer David Hendin recently decried what he called "the shrill voices that sometimes get more credence than they deserve." He wrote in *Quill Magazine* about the environmental doomsayers and how they have sometimes made things terribly difficult for reporters to untangle. Hendin told about a symposium he attended in New York City on urban biology. One of the speakers said this: "The cost of one B-52 bomber could pay to install *all* the necessary sewage treatment plants in New York City four times over."

The audience – mostly environmentally active college students – responded with applause and cheers. The next speaker was a man who at that time was New York City's environmental protection chief, Dr. Merrill Eisenbud.

Dr. Eisenbud stood up and challenged the speaker, asking if he knew just how much a B-52 bomber costs. The speaker responded by saying he wasn't sure, but he remembered a figure of $12 million. And what about the cost of a sewage treatment plant in New York, asked Dr. Eisenbud? The speaker estimated about $4 million. Eisenbud quickly informed the speaker that New York was completing a sewage plant at that time, which would cost *$220* million.

It wasn't that the speaker was deliberately trying to mislead the audience. He just didn't know what he was talking about, a disease which is guaranteed to strike many of us. The scientific community can help by keeping us posted on many things.

Author Hendin points out that despite all the fuss created by the pollution cleanup campaign, it's hard even to define some environmental problems. To a person living in the inner city, he points out, environment is a leaky roof, rats in the kitchen, and peeling, lead-based paint which sickens children who eat it.

These are the things, it seems to me, to which the modern-day scientist must address himself. He has an obligation, I think, to keep himself well informed on a variety of things – national and international – so he can conduct his scientific efforts not in that ivory tower but in the context of a real-world situation, as a member of the human race as well as of the scientific community.

The task is a big one. There is no doubt that (as Dr. Philip Abelson has pointed out) there has been a rise in anti-science attitudes. Perhaps it's because (since the Sputnik era) science has been over-sold to the American public. Perhaps it's because science is still groping for an adequate method of communicating with the American public. At the height of the manned space program, almost anyone on the street could have told you that the Apollo booster was fueled by gases in liquid form, that they had to be kept super-cold, and that sections of the booster fell away as the spacecraft reached certain altitudes. They could tell you a number of the exercises the astronauts went through in getting in and out of the lunar module on the moon and what they did while walking around on the surface of the moon. (Their information came through a massive public education program. Some have called it propaganda.)

But ask the average non-science-oriented person about some of the major areas of current research and see what blank stares you'll get: What, for example, is magnetohydrodynamics and what can it mean for us? What are free atoms – and why? What's the significance of the theory of continental drift? What are quasars and how do they figure into the overall scheme of things? What can the study of oceanography do for us? Getting back to the moon ... has the man on the street been told in a meaningful way *why* we keep spending his money to go *back* to the moon?

Environmental matters may be easier to explain. Ecological problems are often those that can be seen, touched, and in some cases *smelled*. The link between a polluted stream and what causes it doesn't take too many intermediate steps, if it's handled right. Environmental indices – made public – will help.

At the risk of incurring the wrath of those in the science community who feel what they are doing should be in the realm of a much-guarded secret, may I, in conclusion, make a plea for more open-ness, more communication? Surely it's time for men of science to shoulder their burden as *local* residents and *world* citizens. Surely it's time, for example, to come up with a list of environmental priorities for each section of the country, a list to which the general public can relate. Surely it's time for members of the environmental science community to come forward with environmental indices and in fact to do their utmost to explain their ecological concerns to the public at large in ways that make sense. And without looking down their scientific noses.

When they do, I know they'll get a new feeling of having really helped their fellow men.

And not just *incidentally,* they will have helped make my life as a reporter that much more meaningful.

EVALUATION OF NATURAL ENVIRONMENTS

Martin Murie

Member, Environmental Studies Center
Antioch College
Yellow Springs, Ohio 45387

My main concern in this report is to suggest a rapprochement with nature in which nature is respected but we people are not required to hate our presence on the planet. I define "natural environment" as any place where organisms exist in mutual relations that are free of direct management or other drastic human intervention; but presence or absence of people is not a criterion. Plant succession is a key concept for recognizing the vast number and variety of natural areas that are usually not considered natural simply because humans are nearby. With such an attitude, we are free to set up methods for evaluation that are realistic. Two broad categories of environmental traits important for human enjoyment are proposed. These are *variety* and *accessibility*. These categories are supported mainly by examples of exploratory evaluations of diverse bits of habitat in the United States, England, Switzerland, and Yugoslavia. Conventional methods are also discussed, and some descriptions of alternatives to these are presented, particular reference being made to the work of Luna Leopold[2] and Ian McHarg.[3]

DEFINITIONS

In biological affairs, in any terrestrial situation, a process called succession is nearly ubiquitous. This process is a change in species composition as the organisms themselves change their own environment, and it is present wherever the so-called climax equilibrium is absent or wherever human counter-action is not rigorously in force. It is a constant tendency that lurks about all of our works, a major thrust of nature. If we simply say that the degree to which succession prevails is the degree to which an environment is natural, then we have an objective measure of the extent of human intervention. This measure is a crude scale, or maybe spectrum is a better term. At one extreme is the undisturbed plant and animal community, which is in some stage of succession or is in the stable, climatic climax stage. This could be called wilderness. At the opposite is complete management or destruction of organic life by humans. This latter, zero end of the spectrum includes the thoroughly urban habitats and areas of very intensive agriculture. Other examples of this extreme can easily be imagined: airport runways, space stations, etc.

To discuss the use of this scale, I will refer to a study[6] of the Little Miami River of Ohio and the Green River of Wyoming. In this study, we came to assume that any stage of succession is as natural as any other, but we also recognized that people of our culture often place greater value on those areas that are at or near the presumed primeval, primitive, or climax stage. To take this bias into account, we distinguished and labelled sectors of the two rivers, using the single criterion that the most natural environment is that most similar to our theoretical reconstruction of pre-settlement flora. The same was done with the faunas. Comparison of 18th and 20th century bird and mammal faunas was interesting. It revealed that only nine mammal species and five bird species are extinct in the Little Miami valley area. The Green River area has three or four mammalian extinctions.

I would now modify this approach. I think that much greater emphasis should be given to the degree to which the full potential of *current* communities is approached, including the potential of a given vegetation complex to support a given set of animal associates, and less emphasis on the primeval past as a standard of comparison. Figure 1 illustrates this point.

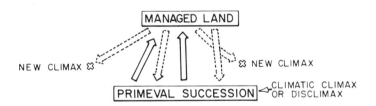

Fig. 1. Changes in status of land units.

I read this figure as follows: All arrows directed downward are escapes from the managed land categories, and upward arrows represent alteration of primeval conditions by man. Communities developing after escape from human domination might return, eventually, to the original community composition, or they might develop a different stability. This latter outcome could just as well be called a type of disclimax. The important point is that we can think of many degrees of naturalness developing along the paths of these downward arrows. We thus avoid a yes-or-no attitude, and this is consonant with the gradual, changeful process of real succession.

We might wish to further complicate our classification system by making a distinction between succession in places whose substrate has not been completely changed or removed by man and succession in areas whose substrate is man-made. When castle walls or a mine tailing move toward forest or perhaps some other final form, we could call it "succession from a man-made base." There are many fascinating examples of these kinds of domination by plants; in fact, their study would make an excellent hobby. However, it would not be necessary to a successful classification scheme to separate man-made bases from others, and in some cases it might be too laborious to do so. The only essential, in my view, is to accept the successional process as one that is changing the former man-made environment into a new and fully natural one.

THE PROBLEM OF PREMATURE JUDGMENT

In pioneer models of evaluation methods, such as that of the Craigheads,[1] careful personal judgments were made on relative ranking of features of environment that were taken to be the key items relevant to the basic categories of outdoor recreation. These basic categories are a common enough list, known to all of us: swimming, boating, canoeing, fishing, camping, hiking, hunting, enjoyment of scenic beauty. No matter how well this evaluation is done, the fact remains that the evaluator has a very heavy burden of responsibility and power. He is both judge and jury, both fact-finder and interpreter. We have to take the word of the evaluator because a system for presentation of his basic data is not available. Sometimes it is clear that there are no data in the common sense of that word; rather, the judgments are made quite offhand. I want to risk quoting out of context in order to give you an example of this practice. In an official study[4] of part of the Green River of Wyoming, the "scenery" of a large section of the river valley was judged to be "not inspiring," and factors limiting the value of this section for "hiking," "nature study," and "sightseeing" included "lack of tree cover" and "plain open country scenery." I hope you agree that a large piece of potentially valuable land should not be passed over by use of such unsupported terms.

My basic assumption is that the evaluator should be a purveyor of information to a wide circle of other interested parties who will make the fundamental decisions. We need, as evaluators, to find some objective procedures that will prevent our making premature decisions. Certainly the data gathered should be relevant to human interest and use of the area, but initially, they should be cast in a form that is more general, neutral, and informative than the approximately eight outdoor recreation categories (fishing, camping, etc.) that I have already listed.

The fault of current methods has been known for some time, and a number of investigators have offered alternatives. I will describe my efforts, which are strictly confined to biological factors, and will later comment on other work. The categories of environmental resources I finally selected for exploration are only two: variety and accessibility.

VARIETY

Diversity and complexity are basic characteristics of life because evolution of species and communities is based on selection of variants, leading to increased complexity. Given these as fundamental in our ancestral background, and in our present habitat, it seems that delight in variety as part of our perception of nature is not merely a modern seeking for titillation. We can accept it as a positive good. Since plant life is ecologically prior to animal life, we can begin with a look at plant variety.

Some of the ways that this presents itself to us are: species composition, species assemblages, and physical structure of the vegetation. In a survey of a million square yards of terrain near the mouth of the river Ouse, near King's Lynn, East Anglia, I discovered that even an apparently uninteresting piece of land has unsuspected features when one tries to describe and map all of it. The vegetation assemblages (stands) were mainly various types of hedges, extensive cropland, and some pasture. There were also a grassed dike, roadside verges, and a bit of wasteland supporting trees and shrubs. Each stand was labelled with the name of its one, two, or, at most, three most dominant species. I also noted the number and kinds of vertical layers in each stand, using a bare minimum of three types for the terrestrial stands: grasses and forbs, shrubs, trees. These layers could occur in one to seven different combinations. I also divided aquatic vegetation into three layer types: bottom-hugging, submergent, and emergent growth. Stands were taken to be different from each other if there was a change in either a dominant or in the layer combination. For example, a hedge of brambles, hawthorne, and grass was considered different from a grove of brambles, hawthorne, oak, and maple. Areal coverage of each stand was measured, and the proportion of total area occupied by each was calculated. This survey showed that the dominant plant assemblage is cropland (91%), that only three different layer combinations are present in the terrestrial parts and one only in the aquatic parts, that in a total of 21 stands there are nine different kinds of stand, and that there is a tiny enclave (1.3%) of un-managed land. The species composition is hinted at in the names of the nine different stands, and all dominant plants and their distributions and combinations are also readily available in the list of stands and the accompanying map. Other excellent, though more laborious survey systems of this type are available, e.g., Ohman and Ream.[7]

For comparison of different areas, the data would be most accurate when drawn from relevant parts of one region. In East Anglia, for example, the proportions of one-, two-, and three-layered hedges can be interpreted with fair confidence because the layers are consonant in their species composition. In contrast, species composition of hedges in Galloway or Cumberland are different; comparison with those in East Anglia would be correspondingly difficult.

My next example, from an alpine meadow on Alp Magriel, Disentis, Switzerland, is a comparison of two very similar plots. These are really components of a single meadow, only a few yards apart. They had nearly all of their plant species in common. The dominant vegetation in both was alpine turf. Various combinations of herbs, grasses, low shrubs, and rocks bearing lichens formed small islets in the turf. The main difference was that one of the plots had a much larger number of vegetation stands (93 vs 58). Only two of the twelve different kinds of stands were not held in common by both plots. The plot having the largest total of stands also had the greater proportion of its area in the grass, forb—shrub layer combination (13.8% vs 7.3%), a lesser proportion in the grass—forb layer type (turf). This example shows that when areas compared share similar climate, substrate, and regional flora, the differences located by the plant variety census can be quite subtle, but also reliably quantified. Fortunately, many of the controversies in conservation and resource planning involve decisions about different parts of a single area.

For a more complicated case, here are a few of the observations made on the Little Miami River. As one travels up or down a river system, one of the basic kinds of changes is the termination of species distributions and the beginnings of other distributions. These changes can be gradual or abrupt. Referring to this type of change as Distribution Change (DC), and scoring each of our sampling stations for the number of plant associations beginning or ending at that station, we can plot the total shift of plant distributions throughout the length of the river valley. It is a measure of the extent to which the variety of plant life is spread out, distributed, in the river system. For example, in Fig. 2, the DC scores of all sampling stations on the river are plotted. The data are for the main stem only. Also in Fig. 2 is a plot of a measure we called Inter-Association Change (IA). This is simply a sum of the total number of plant associations at a given station that differ from the two neighboring stations. The kind of plant distribution that contributes most heavily to this score is that in which associations are numerous and discontinuous; a plant association that is continuous throughout the area would have no chance of contributing to the measure. Unique stands would also be counted but would not be singled out in any special way. The IA score is an attempt to locate segments of the valley where vegetation combinations change most often.

An official study[8] of the Little Miami resulted in a classification of the river into three categories: "Wild," "Scenic Recreation," and "Urban." These sectors are labelled in Fig. 2. It is interesting to note that the two variety scores just described are highest in a part of the river not encompassed by the "wild" designation; that a part of the river not included in the official study, the uppermost reaches of the headwaters, shows a respectable score; and that the "urban" sector, especially near Cincinnati, scored relatively high on both measures. A similar lack of congruence between variety scores and the part generally considered most valuable occurred on the Green River. There the lower Green surpassed the upper Green in vegetational complexity. I conclude from these results that neither the proximity of a major urban area, e.g., Cincinnati, nor the impression of an area as being scenically inferior are necessarily bars to vegetation diversity.

In the city of Mostar, Yugoslavia, is an isolated hill, part of which is a formal memorial to the Partisans and a greater part of which is a largely untended complex of old orchards, overgrown walls, pines, cypruses, and other plants. In a brief, crude survey of this part of the hill, I counted 15 different vegetation stands. Only 2.3% of the area was, in my judgment, managed. Thus, in the midst

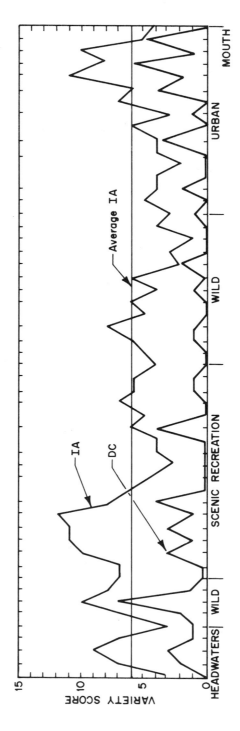

Fig. 2. Vegetation variety scores along the **Little Miami River, Ohio.** DC = distribution change; IA = inter-association diversity. The recreation use categories developed by the Ohio Department of Natural Resources are shown on the abscissa for the river, from its headwaters to its mouth.

of the urban scene, surrounded by streets and housing, an integral island of natural area exists. Mapping the biotic patterns of such outposts might lead to discovery of new qualities, the technical rigamarole of plant census and map-making being aids to insure that we see more in the environment than merely what we expect to see.

ACCESSIBILITY

Whereas variety can be thought of as a wide category of indices to what is present in the biotic system, accessibility implies those factors affecting people's chances of making contact with what is present. Besides the obvious factors, such as trails and roads and legal rights, there are the more strictly biological constraints and opportunities. Angling and its long history of lore, expertise, and popularity is a good model of this sort of thing. Fishermen have evaluated streams, lakes, and salt water for generations in terms of contact between fish and fisherman. Frequency of pools and riffles, food supply, water depth and turbidity, and so on constitute these factors. Similarly, there are conditions that enhance or inhibit human contact with other kinds of animals, but, just as there should be room for fish-watching, scuba-diving, etc., in addition to angling proper, so we should look for all sorts of animal–human interactions in addition to hunting and the other standard sports.

On the Green River, we tried to simulate some of these various activities geared to contact between mammals and humans. Some of the relevant features of mammalian life are the nocturnality or diurnality of each species, because humans are basically diurnal; and the degree to which a species of mammal makes tracks or other sign. Thus, a woodrat is seen less often than the diurnal ground squirrel or prairie dog, and deer mice can be tracked much more often than jumping mice. In addition, the rarity of each species in a given area is important. Finally, there are some species of mammals that usually must be trapped or shot in order to be identified. Many bats, at least in our present state of skill with bats, cannot be identified on the wing. Some rodents, insectivores, and a few others are difficult or impossible to identify in the wild. This is both a challenge to mammal watchers and a useful warning that a rich list of species on paper does not guarantee complete acquaintance in the field.

Birds have a different set of conditions affecting our rapport with them. Unlike mammals, there are few strictly nocturnal birds; so this characteristic is less relevant. Neither are tracks relevant, in general, although in special situations, such as shorebird environments, tracks can be important. On the Little Miami, we arranged our bird list according to residential status, which of course has been done for years by ornithologists, and we indicated those species that are uncommon. These are no improvement on existing ways of guiding people to birds; we recognize that bird-watching has a long and painstaking history. There is room for improvement, but right now the sophisticated state of bird accessibility can serve as a model of what might be done with other more neglected kinds of life.

Any irregularities in distribution of animals in an environment contributes to internal variability. This is also a factor in accessibility. During the study of the Little Miami, birds identified were recorded according to location. An interesting result was that the 29 resident birds judged to be common or easy to find were distributed throughout the river valley, whereas among the 24 species that were less common or harder to find, nearly two-thirds were noted in the northern half of the valley. This finding is not conclusive because the listing of species was not done in a rigorously controlled way. However, a properly constructed distribution pattern could be a useful guide to the chances of encountering each species.

A similar pattern turned up on the Green River mammal study. The less frequently observed animals were nearly all recorded in the upper Green or the Big Sandy tributary; and few of these were on the lower Green.

As every deer hunter knows, the type of habitat has a great bearing on hunting techniques and the difficulty of the hunt. By analogy, the habitats of many kinds of animals are important to many kinds of man—animal interactions. We should no longer be content to designate any area that holds animals as prime land for enjoyment of wildlife. I would guess that there are large tracts of wilderness that, in spite of their irreplaceable value as natural ecosystem, are far inferior to some suburban and farm areas in wildlife accessibility. Animals do not draw boundaries in the same way we do.

CONCLUDING REMARKS

To my knowledge, two of the most sophisticated models for environmental evaluation are those of Luna Leopold[2] and Ian McHarg.[3] Leopold begins by choosing a fairly large list of environmental traits that can be ranked according to strictly objective criteria. Among physical factors, for example, are the width of a valley and height of adjacent hills; among biotic and water quality factors, there are turbidity, amount and kind of plants, etc.; and "human use and interest" items include presence of trash, accessibility, land use, and view confinement. One of the special and promising features of Leopold's scheme is a method of comparing any measured item according to relative uniqueness among the total collection of areas evaluated. Another interesting feature is the method of finding composite scores for two or more items. For example, river depth and river width are plotted together as two variables; this plot is projected onto another line which becomes the ordinate for a second graph whose abscissa is the frequency of rapids and falls. The resulting graph gives a spectrum called "Scale of River Character," showing relative positions of each river, with "small and placid" at one extreme and "large and rapid" at the other. This is an intriguing way of handling large amounts of data without obscuring basic findings in the depths of a computer. It becomes debatable when the composite values achieve esthetic status. Leopold's example of this is a complex of graphs, laid out just as in the preceding example. A scale of values, "Landscape Interest," is generated by combining rankings of "width of valley," "height of nearby hills," and "scenic outlook." The latter is a measure of the degree to which views are confined. "Landscape Interest" is a scale with "spectacular" at one extreme and "ordinary" at the other. This scale in turn is combined with relative urbanization to yield a final measure, "Scale of Valley Character," ranging from "spectacular and wild" to "ordinary and urban." I cannot do justice here to the details of the generation of these scales. My description is sufficient, however, to point to the debatable part of this approach, namely, physical aspects of environment are combined to produce esthetic judgments. This crossing from the level of physical measures to human perception can be negotiated only with agreed-on rules for mapping of one set of variables on another set. It is true that, as Morisawa,[5] Shafer et al.,[9] and others have shown, there is an amazing amount of agreement among people as to their preferences in the realm of visual enjoyment of natural environment. But is that sufficient? I should think that the next steps are to determine to what extent our scenic pleasures are the joys of recognition and whether current fashion in scenery is just that — fashion. Also, possible minority views, i.e., alternative perceptions, should be looked for.

I have a similar cautionary feeling in relation to some aspects of Ian McHarg's proposals. Certainly if regional planning is the way of the future, McHarg's in-depth probing of values that must be dealt with is a fine beginning. One of the hallmarks of his work is the insistence that intrinsic capabilities and limitations of resources should be seriously recognized in order to work out ways in which human structures and actions can interlock with these values rather than destroy them. In this context, plant and animal communities, wildlife, outdoor recreation lands, and, of course, "scenic beauty" are given important places. These resources are apparently to be recognized and allocated by means of appropriate study and surveys. But what special techniques are in fact appropriate for deciding on the scenic quality of competing sections of landscape, or on why certain areas are better than others for recreation? Do we all have an underlying agreement on these matters? I have the impression that McHarg has a strong optimism about the emergence of correct decisions once patterns of nature are thoroughly gone into by man in search of harmonious and meaningful relation to nature. How can anyone disagree that the most important thing we can do in relation to nature is to understand and try to fit in? However, we are a species, too, with strong traits and powerful histories. Among these traits are our preoccupation with something called "beauty" and capacity for joy in being here and in being active in our habitats. I don't see any science by which decisions in these realms can be easily or casually generated. My own suggestions about evaluation of natural environment do not offer a direct route either. Rather, I see such methods as being useful only when operated within a wide context where there are various stages in the evaluation process and also a provision for argumentative and controversial stages, all of these taking place *before* any plan or proposal is considered even tentatively complete. The few ways of evaluation that I have presented are an approach that aims to complicate the process in a useful way.

ACKNOWLEDGEMENTS

The field work and data analysis in the biological part of the river studies quoted in this paper were carried out with major assistance by the following students of Antioch College: Chris Chennel, aquatic survey; Dean Denno and Bonner McAllester, bird and mammal surveys; Rudolph Darling, bird and vegetation surveys.

REFERENCES

1. Craighead, F. C., Jr., and J. J. Craighead. River Systems. Recreational classification, inventory and evaluation, *Naturalist.* No. 2: 3–13 (1962).
2. Leopold, L. B. 1969. Quantitative Comparison of Some Aesthetic Factors Among Rivers. Geological Survey Circular 620, 1969.
3. McHarg, I. L. *Design With Nature.* The Natural History Press, New York, 1969.
4. Mid-Continent Regional Study Team. Wild Rivers Study; Upper Green River, Wyoming (Fontanelle Reservoir to Green River Lakes): Preliminary report. (mimeographed) U.S. Depts. of Interior and Agriculture, 1963.
5. Morisawa, M. Evaluation of Natural River Environments, Phase II. U.S. Dept. of the Interior, Office of Water Resources Research, Project No. C-1779, 1971.
6. Morisawa, M., and M. Murie. Evaluation of Natural River Environments, U.S. Dept. of the Interior, Office of Water Resources Research, Project No. C-1314, 1969.
7. Ohmann, L. F., and R. R. Ream. Wilderness Ecology: a method of sampling and summarizing data for plant community classification. U.S. Dept. of Agriculture, Forest Service Research Paper NC-49, 1971.
8. Price, W. The Little Miami of Ohio: A study of a Wild and Scenic River. Ohio Dept. of Natural Resources, 1967.
9. Shafer, E. L., Jr., J. E. Hamilton, Jr., and E. A. Schmidt. Natural Landscape Preferences: A Predictive Model, *Journal of Leisure Research* 1(1): 1–19 (1969).

BIBLIOGRAPHY

Appleyard, D., K. Lynch, and J. R. Myer. *The View From the Road.* Massachusetts Institute of Technology Press, Cambridge, Mass., 1964.
Barker, R. G., and L. S. Barker. The psychological ecology of old people in midwest Kansas and Yoredale, Yorkshire, *J. of Gerontology* 16(2): 144–149 (1961).
Clark, R. K. Before the Picturesque. *Landscape* 17: 18–21 (1968).

Errington, P. L. Of Wilderness and Wolves. *The Living Wilderness* 33: 3—7 (1969).

Kates, R. The pursuit of beauty in the environment. *Landscape* 16(2): 21—25 (1966).

Lowenthal, D., and H. C. Prince. The English landscape. *The Geographical Review* 54(3): 309—346 (1964).

——————————· The American scene. *The Geographical Review* 58: 61—88 (1968).

Lucas, R. C. Wilderness perception and use: the example of the Boundary Waters Canoe Area. *Natural Resources Journal* 3(3): 394—411 (1964).

MacKaye, B. *Expedition Nine. A Return to a Region.* The Wilderness Society, Washington, D. C., 1969.

Nash, R. *Wilderness and the American Mind.* Yale Univ. Press, New Haven, Conn., 1967.

Sargent, F. O. Scenery Classification. Agricultural Experiment Station. Univ. of Vermont, Burlington, 1967.

Saarinen, T. F. Perception of Environment. Resource Paper No. 5. Commission on College Geography. Assoc. of American Geographers, Washington, D. C., 1969.

Shafer, E. L., Jr., and R. C. Thompson. Models that describe use of Adirondack campgrounds, *Forest Science* 14(4): 383—391 (1968).

Sonnenfeld, J. Variable values in space and landscape: an inquiry into the nature of environmental necessity, *J. of Social Issues* 23(4): 71—82 (1966).

Yi—Fu Tuan. Attitudes toward environment: themes and approaches, pp. 4—17 in *Environmental Perception and Environment.* Research Paper No. 109, Univ. of Chicago, Dept. of Geography, 1967.

INDICATORS OF ENVIRONMENTAL QUALITY OF URBAN LIFE: ECONOMIC, SPATIAL, SOCIAL, AND POLITICAL FACTORS

Chester Rapkin

Professor of Urban Planning and Director, Institute of Urban Environment Columbia University New York, New York 10027

Robert W. Ponte

Assistant Professor of Urban Studies Barnard College New York, New York 10027

THE NEED FOR URBAN INDICATORS

In a nation that is turning increasingly urban, the quality of the urban environment constitutes a major test of the level of its well-being as a society. The 243 metropolitan areas of the United States now comprise 69% of the nation's population and are more than ever the center of its economic and cultural life. Although planners and social scientists have long questioned the low priority given to enhancing the urban environment, growing segments of the public share their concern, recognizing glaring deficiencies in their surroundings. The national discussion about priorities in government expenditures is clearly tied to this realization and seeks to redirect government effort toward such pressing areas of domestic need as the improvement of urban development and housing, the health of the people, crime prevention, recreation and education, and environmental pollution, among others.

Without a systematic program to eliminate our slums and reconstitute our cities, the urban ambience has and will continue to lag behind less important components in the social system. Clearly, there is no question among informed persons that such a program is urgently required; rather, the current dilemma is one of formulating an effective and workable program that will contribute to the meeting of social needs. Even though political forces show signs of turning public policy in this direction, government planners lack tools for measuring the magnitude of public needs, as well as their changing characteristics over time.

At present, it is difficult to judge which problem areas are worsening, which are stable, and which are correcting themselves through regulation or through the market process; and it is thus difficult to prescribe the needed level of government intervention for each. Information is also lacking on the relationship among the different problem areas and the likely effects that a particular treatment of one area may have on the others. In short, there is a need for more precise measures depicting social needs, indicating their direction of change, and evaluating the consequences of different types and degrees of intervention.

To be useful to policy-makers, however, these measures must be susceptible to expression in terms of explicitly stated social goals. Measuring the gap between the present level of well-being and the national goal in each problem area would be the preliminary purpose of government efforts. Regular readings of the extent to which the gap is being reduced could serve to evaluate the effectiveness of government programs. It has been particularly difficult to obtain a reasonable degree of agreement on national social goals because of wide divergence in views among the various economic, political, and class groups in our society. In the most part, we have not been successful in moving beyond general statements to quantified standards,[10] except in certain specialized fields like housing, air pollution, and open space.

Economic indicators, developed by social scientists during the 1930's, offer a basis for conceptualizing measures of one dimension of urban environmental quality. These statistics have been refined, so that we now have accurate rates of employment, levels of living costs, and industrial production totals for the nation and for metropolitan areas. Census data furnish corresponding information on the demographic and housing aspects of environmental indicators. Nevertheless, these social statistics cannot as yet be combined into a consolidated system of accounts as can economic data. National income accounts summarize the status of the economy by aggregating related data into the gross national product, revealing whether the economy has grown or declined in balance. The many components of national income can be combined by utilizing a basic unit of measure – the dollar; and even though a vast variety of phenomena are incorporated, the framework is sensitive enough to reveal even a mild contraction or a modest advance.[21] Such measurements as rates of growth or decline allow analysts to assess the effectiveness of particular policies.

For all their usefulness, however, national or regional income accounts are by no means complete measures of the status of the urban environment. They do not include the social costs of production; and in fact by adding all goods and services together, actually consider both the products of a smoke-spewing factory and the resultant expense of combating soot in the adjacent areas as contributions to national output. More importantly, national income figures touch only tangentially on most of the essential elements of an adequate urban environment, such as decent shelter, suitable community facilities and services, family stability, personal security from crime, enriching educational and cultural opportunities, swift and safe transportation, responsive government subject to local participation in decision-making, and an atmosphere of social justice.

A system of social accounts that combines these factors in a significant manner is still a distant prospect. Many of the statistics necessary for such a tally are gathered only when required by the administrative needs of agencies; and some areas of concern, such as alienation, may not even lend themselves to statistical measurement consistent with respect for privacy.[28] Moreover, a common unit of measurement is lacking among social indicators. Values play a crucial role in determining the quality of life, and measures cannot be uniformly reduced to a unit of currency, as in the case of national income statistics. It should be recognized, however, that not even all economic indicators are additive, and some must be evaluated separately because they are measured in other types of units, e.g., percentage of labor force unemployed or proportion of households earning less than half the median income. An added complication is the fact that useful statistics for social indicators are available from a wide variety of sources at varying intervals, rendering it extremely difficult to put together a comprehensive set of indicators on a regular basis. There is little in the theory of social science to indicate even a desirable frequency with which structural changes in the social pattern should be observed. Finally, social indicators still lack an organizing concept expressing basic cause and effect relationships.[23] Without a complete understanding of the environmental system in cities, including the indirect effects of individual social forces, discrete indicators cannot be weighted and added together to yield an aggregate index of urban quality. Rather than seek the unattainable goal of a perfect reading of social conditions, it would be well to adopt an empirical attitude, seeking to refine existing measures and to understand their implications better, while devising new proxy indicators of environmental factors not presently quantifiable.

Most of these observations would apply to social indicators, whether nationwide in scale or merely concerned with a single metropolitan area. It is important, however, to understand that urban indicators do have some basic differences from national social indicators.[9] Because they pertain to built-up areas limited in size, they must reflect the needs of a highly interdependent population. In general, urban dwellers are much more dependent on public services and facilities than are the rural and small-town populations. The demands and strains of high density living in a climate of intense competition also severely test family stability and the psychological balance of individuals. The pace of social change is quickest in cities, and traditional forms of social organization tend to break down faster, leaving people uncertain of acceptable norms. Social pathology is the by-product of this process; and demands upon the schools, the courts, the police, the hospitals, and the welfare groups increase as people turn to impersonal public agencies for functions formerly performed by themselves, their families, or their neighbors. Thus, urban indicators must cover many areas, such as health care, crime prevention, cultural level, minority status, political participation, transportation, housing adequacy, land planning, and aesthetic satisfaction.

A fundamental purpose of a system of measures of urban environmental quality is to make comparisons among metropolitan areas possible. Each metropolitan area should be evaluated in the same categories of measurement, employing the same units as all others. The result would be an array of numbers — one for each category — that, while not subject to aggregation, could be compared category-by-category with all other metropolitan areas. A recent publication[13] of the Urban Institute has adopted this approach in analyzing the relative status of metropolitan Washington, D.C.

Urban indicators should be oriented to public policy questions. As comparative measures, their most basic use will be to guide the allocation of federal assistance funds among metropolitan areas. For the first time, it will be possible to employ objective criteria indicating the relative need of regions in each policy area. This use of urban indicators takes on special importance as the federal government considers the form which revenue sharing should take. It should be recognized, however, that even though the distribution of federal funds is increasingly becoming subject to metropolitan planning considerations, most assistance is allocated directly to state and local governments, rather than to metropolitan-wide entities. This means that sub-metropolitan indicators are needed as well, so that the central city and major suburbs can be evaluated vis-a-vis their competitors for federal funds across the country.[2] And, as Perloff[23] has pointed out, the indicators should also distinguish between residential and nonresidential neighborhoods and among types of communities like ghetto areas and middle class neighborhoods.[4] Acquiring complete data for such areas of limited size is extremely time-consuming and expensive, except when census tract information happens to coincide with the area in question. But at the scale of the central city or major suburb, such data might not be any more difficult to obtain than for the metropolitan area as a whole; in fact, the data needs of municipal agencies may generate more local data than are available on a metropolitan scale.

As indicators for each metropolitan area become available for different years, they will also be useful in disclosing which regions are meeting their problems successfully, which are stumbling along, and which are not making any headway. Government will be able to evaluate the effectiveness of its programs by checking its rate of progress in solving problems in various policy areas. Local effort will, of course, play a large part in determining the success or failure of government initiatives, and federal decision-makers will want to adjust grant-in-aid payments so that there is a maximum utilization of local resources, even if the absolute need for environmental improvement is great. Demonstrated progress will require a careful federal evaluation of the need for additional funds, depending upon the amount of local momentum and commitment, in addition to the relative status of the different policy areas and their effect upon each other. Lack of progress or retrogression may signal the need for additional federal assistance or may indicate local disinterest or absence of effort. Clearly, simple measures of need will aid in the allocation of federal assistance, but they can never constitute more than a portion of the decision process when so many difficult-to-gage variables are involved and when the allocation of huge sums of public funds is at stake.

THE SOCIAL AND BEHAVIORAL CONTEXT

The scope of this paper differs from other discussions of the urban environment inasmuch as it is not concerned with the quality of such profoundly basic elements as water, air, and sound. Contamination of water and air has a deleterious effect upon the physical health of the individual and, if severe enough, can be lethal. Their effect is direct, indisputable, and universally accepted. The problems of noise contamination are somewhat different, being more in the realm of psychology and aesthetics than the other two but also having direct physical effects. Once the intensity of the sound waves goes beyond the tolerance threshold, the function and structure of the hearing mechanism is affected and psychological disturbances are generated. In the case of all three, the problem of quality is concerned with the physical character and nature of the elements and the possibility of adverse consequences largely on the physical health of individuals in the community. It is certainly true that society would cease functioning abruptly if air and water became too contaminated for ready use. But long before this would happen, its consequences on the physical health of individuals* would have become the overwhelming preoccupation of mankind.[5]

The elements with which we are concerned in this paper are also, but not entirely, physical in nature. Rather than the natural components of the environment, they involve man-made artifacts — the homes, work places, public buildings, recreation areas, open spaces, and streets and highways — and the arrangement of all of these components within the urban pattern. But in addition to the physical elements, the urban environment also subsumes a far more significant social component, having to do with the way in which men have organized themselves to live in large numbers in densely concentrated geographic areas. Just as the urban physical structure has abundant variations, so does the social organization within which urban dwellers function.

How much of the social system is attributable to the entire society of which the urban area is a part and how much of it is a function of the city itself is difficult to determine. With time, this distinction diminishes in importance as world society becomes increasingly urbanized. What is important to explore is the extent to which the quality of the urban environment (now defined to exclude problems of contamination of water, air, and sound) influences health, behavior, and personality. Proshansky et al.[24] have summarized the two diametrically opposite psychological approaches that have sought to explain the relationship between the individual and his environment. The objective approach, epitomized by Watsonian behaviorism and developed in the more recent work by Skinner,[29] holds that the individual responds in a fixed and

---

*G. L. Engle[5] has classified the external injuries that strain the individual into four general categories: physical injury; insufficient food, water, air, etc.; infection by pathogenic organisms; and psychological stress.

discoverable manner to identifiable stimuli. In contrast, the phenomenological approach, developed by Kaffka and expanded by Lewin, holds that behavior is not a response to the stimuli of the objective world but issues from that world transformed into an inner world by an inherently cognizing organism. Building on these extreme views, Proshansky et al.[24] assert that an individual's perceptions and behavior are not independent of the social milieu of which he is a part. The authors maintain that "from a theoretical point of view, there is no physical environment apart from human experience and social organization ... and the individual's response to his physical world is never determined solely by the properties of the space and events that define it ..." They illustrate this point by explaining that "crowding," for example, is not simply a matter of the number of persons in a given space. Being crowded depends on many factors, including the activity going on, the people participating in it, previous experience involving numbers of people, gender of the individual, and the extent to which bodily proximity is acceptable in the given society.

The interaction between the quality of the physical environment and the social context is seen quite clearly if we accept the fact that there would be no housing problem if there were no housing standards. By and large, in the more economically advanced nations of the world, the concern about the quality of housing, work places, and other elements in the urban physical structure stems very little, at this stage of our progress, from the fact that the deficiencies constitute a menace to the physical health of the population. This is not to say that there are no slums in the United States or that rat bite, lead poisoning, and pulmonary diseases are unknown, but rather that we have attained a reasonably high standard of housing and that even slums have a level of facilities and amenities unknown in developing countries or in the western world as recently as a generation or two ago. In essence, our concern about the physical quality of our structures stems from the commentary that it offers on the equity of the distribution of the nation's economic products, on the social symbolism of housing, on matters concerned with privacy, and on amenity.

Housing standards, in effect, represent the community consensus of what each family in a given society should have if larger social goals are to be met. Over time, housing standards have continued to move upward, responding to rising incomes and advancing technology. In this process, the definition of a satisfactory housing unit has moved far beyond the level of concern with health and safety and has concentrated more on convenience and amenity, as basically sound structures became more readily accessible to larger and larger sectors of society.

We would indeed be remiss if we did not attempt to place the effect that environment has on shaping behavior and personality in some larger perspective. What proportion of an individual's make-up is accounted for by external environment generally and by the urban environment specifically? Among the great influences that shape man are his genetic baggage, the family in which he was reared, its composition and the relationships among its members, traumas of various sorts that he may have experienced, biochemical imbalances, or other physical pathologies that influence his life's experiences both inside and outside his family. These experiences include not only interpersonal relationships but also such matters as his social status, country or place of origin, and its cultural heritage and geographic location, as well as his opportunities for growth and development, his actual achievements, and his feelings about them. Placed in this setting, it would seem that the urban environment qua physical structure would have a minor effect indeed as a formative influence on the totality of the individual, an intuition buttressed by the fact that there is considerable similarity in human behavior patterns the world over.

It has been commonplace, if less than scientific, to attribute the stress of modern life in substantial measure to the characteristics of the urban environment. Among the more recent studies that attempt to investigate the validity of this popular belief are explorations by the Regional Plan Association. The RPA maintains that high density has led to superficial, over-routinized experiences and sensory overload, which contributes to fatigue and frustration.[27] This is particularly apparent in "proxemic problems," what Ardrey,[1] Hall,[11] and others have termed "territoriality." They postulate a "body buffer zone," which if violated produces anxiety. In some individuals, the violation of this space causes great agitation. Kinzel[14] has pointed out that many persons imprisoned for crimes of violence are far more sensitive to physical closeness than those jailed for property crimes. The RPA also maintains that density at times stimulates competitive impulses, even among total strangers, which often may be gratuitous or excessive. This, in turn, intensifies the discomfort of being crowded, to which the impulse was an original response.

Urban society has also generated a substantial degree of migration and mobility, which impels a family to find a new home, new place of work, new friends, new shops, new neighborhood, or new schools if it wishes to improve its living or working conditions. For a middle-class family, such movement is dislocating indeed, but for poor families, where the rate of mobility is considerably higher, the dislocation is more profound, mitigated in part by the fact that they frequently tend to stay within the same general community. The typical worker, and more particularly the member of the poorest sector of society, depends upon his house and neighborhood not only for shelter and amenity but also for safety and haven.[7,8,25]

All of these intrinsic characteristics of a large urban concentration tend to generate emotional pressures that frequently rise to intolerable levels and take their toll on the mental health of the urban population. Srole et al.,[30] in a study of the apparently most stable population group of a 200-block area of mid-Manhattan, found that only 19% of this group showed no symptoms of mental illness; 59% gave evidence of mild to moderate mental illness; and 23% showed marked or severe mental illness.

These findings tended to support the suggestions that the tensions of the great metropolis were the root cause for the unbalance, until Leighton et al.[16]

demonstrated that roughly the same incidence was to be found in a small rural area of Nova Scotia. With mental illness so prevalent, one is impelled to ask, "If everybody is crazy, is anybody crazy?" In other words, has pathological behavior become the social norm? Has psychopathology become the accepted order of the day? To state the question in this way makes it sound ridiculous, but it does make one shudder when a prominent psychiatrist[22] states that "the behavior of mental patients varies not in kind but in degree from that of normal individuals."

Psychiatric studies by Faris and Dunham,[6] Hollingshead and Redlich,[12] and Langner and Michael[22] have consistently revealed a high correlation between social class and mental illness, which provides a link between emotional stability and societal form. It is not astonishing to find that the lowest class — deprived of status, recipient of a minimum of the economy's product, and virtually devoid of power — is most susceptible to emotional difficulties leading to mental illness in their struggle for survival in everyday life.

Since each society attempts to produce the the type of person that conforms with its norms, an obvious test of its effectiveness is the degree to which its environmental composition militates for or against its objectives and indeed produces people whose behavior patterns are substantially in accord or at variance with those that are acceptable and desirable. To the extent that the society produces criminality, i.e., presumably volitional antisocial behavior, it has failed in its objectives. To the extent that it produces derangement among its people, that is, involuntary behavior of an unacceptable nature, it has indeed distorted itself violently in a manner that defeats its ultimate purposes. By the same token, a measure of success may be found in the degree to which society has reduced or eliminated the frictions and irritations of daily life, has provided acceptable alternatives to destructively aggressive behavior, and has bred a people capable of coping with the inevitable interactions, stresses, and tensions, as well as the major and minor emergencies that constitute the components of human existence wherever man finds himself.

A PROPOSAL FOR URBAN INDICATORS

Social reporting on the urban environment as a measure of the level of well-being and the effectiveness of public programs is closely related to the systems analysis approach in urban planning. Both use mathematical methods to evaluate progress in moving toward an explicitly stated goal.[21] In systems analysis, outcomes of alternative courses of action are compared in terms of some criteria of desirability, so that the optimum solution can be chosen. The Planning-Programming-Budgeting System (PPBS), the systems technique now used throughout the federal government, first proved itself useful in reorganizing the budget of the Defense Department. Its effective use in the programs of civilian agencies is more limited because federal agencies usually share authority with state and local governments, as well as with the private sector. Rational policy-making for domestic needs requires a deeper analysis of the many forces affecting the course of society, and social indicators are a logical extension of the systems approach. Through the quantification of strategic variables, they attempt to bring a higher degree of precision to policy analysis.

Urban indicators should be conceived in clusters relating to broad policy areas. Not only is it uncomplicated to think of measures according to demographic, economic, spatial, social, and political changes, but indicators for each of these areas carry with them an implicit goal structure. (For a general discussion of an alternative urban indicator system divided along individual, family, and institutional lines, see Moynihan.[26]) While related to each other, the goals for each form natural analytical boundaries. As we discuss each area, its goal orientation will emerge, even though our purpose here is not to set performance targets. Such numerical criteria will have to be determined administratively later, but they will undoubtedly be based upon existing standards in many areas of need. Once indicators for different years become available, the change registered in specific policy areas would provide the basis for a program review in each, directed toward deciding whether resources might be used more productively in other ways. This policy analysis, based upon the demonstrated effectiveness of programs in attaining stated goals, would go beyond the purpose of indicators and would call for a PPBS treatment on a government-wide scale.

We propose, then, that urban indicators be structured in the five major analytical areas mentioned above, so that they are closely tied to substantive problems and avoid merely recounting the expenditures and activities of government.[21] Some of these areas, particularly the demographic and economic, are already subject to sophisticated measures, while others, like social and political indicators, still require more understanding even to devise adequate proxy measures. Our intention here is to summarize the existing level of measurability in each area and suggest new directions for social research in quantifying urban problems.

Demographic Measures

Population counts are the oldest indicator of the state of the nation. They were considered so important that they were required decennially in the original Constitution. The census provides us with basic data on number of persons for the nation as a whole and on age, sex, and race in small subareas.[31] Over the years, the scope of the census has been broadened considerably beyond a simple headcount, and information is now also available on many aspects of demographic change. Rates of national increase (births minus deaths), household formation rates, and the proportion of the population comprised of elderly, minority, and impoverished persons all provide detailed information on the shifting pattern of need in metropolitan areas.

Population growth and decline, as well as changes in its composition, are the most fundamental factors in determining needed government programs and facilites. Demographic data from census to census indicate the magnitude and direction of change and paint the basic picture of need in each region.

Intercensal demographic estimates have become available in recent years, allowing policy-makers even more flexibility in monitoring change. And the expansion of the census into housing, personal income, and business activity has laid the groundwork for measurement in the other proposed policy areas. Nevertheless, additional information is needed on migration flows, neighborhood segregation, and household characteristics.[19] If demographic measures are to maintain currency, more frequent censuses must be undertaken, and it is therefore essential that proposals for a mid-decade census be adopted. If this is not possible, more complete intercensal estimates will be required, especially for subareas of metropolitan regions where rapid change is taking place.

Economic Measures

After population data, economic measures are our most complete component of urban indicators. There are two main categories of economic data that must be considered in rating the well-being of a metropolitan area: the income and assets of the individual, family, and household; and the status of the region's economy.

Income is the most meaningful measure of a person's economic status, and median figures are reported by the census for regions and their subareas. The income distribution, which can be computed from available information, sheds light on the degree of social stratification in an area and the relative hardships experienced by the underprivileged. Data on occupational status are also essential in establishing a region's level of well-being, especially for minority groups. Census data not only permit an examination of status at specific moments in time but also indicate social mobility over time. Employment rates and proportion of households receiving welfare payments are additional important indications of the local economic conditions for which data are available. More complete information, however, is needed to determine the level of personal assets in local areas.

Economic base analysis offers a tested conceptual framework for rating the relative strength of metropolitan areas as productive units. This type of analysis divides the local economy into its constituent parts, disclosing the degree of importance of each and their individual advances or declines. These data are useful in providing measures of the level of employment opportunity present in a metropolitan area and its general growth outlook. A region's economic role in the national scene can also be analyzed from data showing whether it imports or exports capital. Cost of living data are an essential overall evaluation of a region's economic impact on the individual household. Additional study is needed to determine the likely effect on minorities of trends pointed up by these measures.

Spatial Measures

Since measuring of pollution has been more than adequately covered at this conference by others, the spatial components of our proposed urban indicators will concentrate on housing, community facilities, and aesthetic satisfaction.

Housing has both an individual and community dimension. The options of individual households are limited by rent levels, the availability of mortgages, prevailing housing conditions, and their own household size. Their ability to satisfy their needs will depend upon the general availability of housing in the region and the rate of housing production. If the supply is limited, crowding and lack of privacy may result, and people may be forced to settle for housing below the national standard set by the federal government. The census and other data record these characteristics regularly, and their figures would go far in indicating the adequacy of public and private efforts to shelter the population.[26] The prevalence of discrimination remains difficult to gage, however, and more research is required to test such techniques as the "segregation index" proposed by the Taeubers.[32]

Urban planning has devised yardsticks to measure the need for schools, parks, playgrounds, hospitals, museums, and other community facilities.[3] These standards are based upon local population, from the neighborhood to the region-wide scale, but they must be modified according to the special demographic characteristics of each area. For example, an area with an extensive elderly population would not require the typical allocation of playground space but may need a greater than average number of hospital beds. Census data can be used to modify standards for local population needs as subarea goals are set.

The adequacy of local transportation facilities is a major spatial indicator of urban quality. Criteria of measurement are the choice of modes, speed, amenity, and comprehensiveness of coverage. Most of these factors would require the development of new measurements, but one helpful overall indicator might be the average time spent in the journey to work. At the present time, regional transportation studies make it possible to compare the efficiency of local movement systems by utilizing their data on the amount of time spent in the home-to-work trip.

The aesthetic quality of a metropolitan area has only recently become a vital public concern. Design amenities in the townscape are essential to the enrichment of life and serve in themselves to create an awareness of physical beauty. On a more functional level, an orderly environment is necessary if the individual is to be able to orient himself as he moves from place to place in our extensive metropolitan areas. Analytical tools to measure urban design quality are conspicuously lacking, save for intuitive judgments of experts. Lynch, Appleyard, and Myer[17,18] have supplied us with the basic terminology for a more precise system of measurement, but the few existing case studies must be expanded into an extensive body of research if progress is to be made in this area.

Social Measures

Social aspects of the urban environment are perhaps the most difficult factors to measure and most often must be dealt with in terms of their symptoms. For example, rates of mental illness, suicides, and drug addiction are typically used to indicate the prevalence of anomie and inability of individuals to adjust to urban pressures.[13] Public order is usually measured by a variety of crime rates, including robberies. Family stability is determined by the proportion of female-headed households and the rate of divorce and separation. Educational and cultural levels are indicated by census readings of average number of years completed in school and by national educational honors awarded to residents of the area in question. The social status of minorities is rated according to their representation among the high officials of government and business and in the professions. These surrogate indicators may have to serve temporarily until we can better grasp the interrelationships in the social sphere of the urban environment. Intensive research into more sophisticated measures of our social status deserves high priority in the social sciences.

Political Measures

Although in part codified by law, the urban political process is extremely difficult to measure qualitatively, largely because of the considerable amount of discretion given to administrators in determining procedures and priorities. There are four major areas, though, in which we can begin to measure the adequacy of government. First, the responsiveness of government to public needs can be considered a function of its expenditures per capita for a selected list of high priority purposes. Urban governmental effort, mentioned earlier, can be measured by the degree to which municipal income depends on taxation and the extent to which it is supported by grants-in-aid from higher levels of government. Second, public participation in government can be judged by the proportion of the eligible electorate that has registered and voted. In addition, there is a need for a broad-ranging measure of the strengths of citizens groups and of public hearings in the decision-making process. Third, local commitment to social justice is associated with laws forbidding discriminatory practices and the effectiveness of the mechanisms that provide for their enforcement. Fourth, concern as to the future condition of the physical environment can be evaluated according to the quality of the area-wide planning process in the region. If it has an able technical staff supported by strong political backing, there is a good chance that the contribution of the planning process to physical development will be characterized by high standards of spatial arrangement and satisfaction of social needs.

As with the proposed social indices, these political indicators are still many years away from a level of sophistication comparable to demographic and economic measures, but they can be of use by stimulating additional thought on the subject of social reporting. In their estimates of need, they are capable of alerting the general public and its political leadership to the magnitude of the environmental challenge that lies ahead in our cities.

CONCLUSION

At this point in the development of urban indicators, it is still too early to expect an aggregate measure of urban environmental quality. In former years, social scientists were eager to develop the aggregate entity now known as national income, so that a summary statistical measure of the economy would be available. Only after this achievement was accomplished did we realize that the most useful characteristic of the national income account was its components. Readings on the changes in the status of important industries and the components of consumer expenditures became crucial indicators for policy-makers. Similarly, with social indicators we should not be too frustrated by the great difficulty in arriving at an aggregate figure. Some of the measures proposed here have already begun to tell us whether given sectors of the urban environment meet current social needs or whether they are past the possibility of adapting to emerging life styles. As one example, simple figures showing whether or not size and type of households are coordinated with the characteristics of the standing stock of dwelling units can offer an early warning of growing need for investment in housing. In terms of the perceptions to be gained through social reporting, the sum of the parts is clearly much greater than the whole.

REFERENCES

1. Ardrey, R. *The Territorial Imperative*, Dell Publishing Co., New York, 1966.
2. Berenyi, J. (ed.). *The Quality of Life in Urban America, New York City: A Regional and National Comparative Analysis*, 2 vols., Office of the Mayor, New York, 1971.
3. Chapin, F. S., Jr. *Urban Land Use Planning*, University of Illinois Press, Urbana, 1965.
4. Community Renewal Program. *Urban Level-of-Living Index*, Department of City Planning, Pittsburgh, 1965.
5. Engle, G. L. *Psychological Development in Health and Disease*, W. B. Saunders Co., Philadelphia, 1962.
6. Faris, R. E. L., and W. H. Dunham. *Mental Disorders in Urban Areas – An Ecological Study of Schizophrenia and Other Psychoses*, University of Chicago Press, Chicago, 1939.
7. Fried, M. Grieving for a Lost Home, in *The Urban Condition: People and Policy in the Metropolis*, L. J. Duhl (ed.), Basic Books, New York, 1963.
8. Fried, M. Transitional Functions of Working Class Communities: Implications for Forced Relocation, in *Mobility and Mental Health*, M. B. Kantor (ed.) Charles C Thomas, Springfield, Ill., 1963.
9. Glazer, N. Statement, p. 400 in The Quality of Urban Life, Hearings Before the Ad Hoc Subcommittee on Urban Growth, Committee on Banking and

Currency, Congress of the United States, Washington, D.C., November 17, 1969. U. S. Government Printing Office, Washington, D.C., 1969.

10. *Goals for Americans*, Report of the President's Commission on National Goals, Spectrum, Englewood Cliffs, New Jersey, 1960.

11. Hall, E. T. *The Hidden Dimension*, Doubleday and Co., New York, 1966.

12. Hollingshead, A. B., and F. C. Redlich, *Social Class and Mental Illness — A Community Study*, John Wiley and Sons, New York, 1958.

13. Jones, M. V., and M. J. Flax. *The Quality of Life in Metropolitan Washington, D.C.: Some Statistical Benchmarks*, The Urban Institute, Washington, D.C., 1970.

14. Kinzel, A. F. Body-Buffer Zone in Violent Prisoners, presented at the annual meeting of the American Psychiatric Association, May 1969.

15. Langner, T. S., and S. T. Michael. *Life Stress and Mental Health: The Mid-Town Manhattan Study*, Vol. 2, Free Press of Glencoe, New York, 1963.

16. Leighton, D. C., et al., *The Sterling County Study*, Basic Books, New York, 1963.

17. Lynch, K. *The Image of the City*, The M.I.T. Press, Cambridge, Mass., 1960.

18. Lynch, K., D. Appleyard, and J. R. Myer, *The View From the Road*, The M.I.T. Press, Cambridge, Mass., 1964.

19. Milgram, G., and R. W. Ponte. *Demographic Indicators of Metropolitan Problems: A Suggested Framework for the General Guidance of H.U.D. Program Emphasis*, Institute of Urban Environment, Columbia University, New York, October 1969; prepared for the U. S. Department of Housing and Urban Development.

20. Moynihan, D. P. Urban Conditions: General, *The Annals*, 371: 159-177 (May 1967).

21. Olson, M., Jr. The Purpose and Plan of a Social Report, *The Public Interest*, No. 15: 86, (Spring 1969).

22. Page, J. P. (ed.). *Approaches to Psychopathology*, Temple University Publications, Philadelphia, 1966.

23. Perloff, H. S. A Framework for Dealing with the Urban Environment: Introductory Statement, p. 16 in *The Quality of the Urban Environment*, H. S. Perloff (ed.). Resources for the Future, Inc., Washington, D.C., 1969.

24. Proshansky, H. M., W. H. Ittelson, and L. G. Rivlin. The Influence of the Physical Environment on Behavior: Some Basic Assumptions, pp. 8 and 28 in *Environmental Psychology, Man and His Physical Setting*, Holt, Rinehart, and Winston, Inc., New York, 1970. (References to works of Watson, Kaffka, and Lewin are given in this text.)

25. Rainwater, L. Fear and House as Haven in the Lower Class, *AIP Journal*, January 1966.

26. Rapkin, C., and R. W. Ponte. *Indicators of Metropolitan Housing and Residential Construction Needs: A Suggested Framework for the Allocation of Future H.U.D. Assistance*, Institute of Urban Environment, Columbia

University, New York, June 1970; prepared for the U. S. Department of Housing and Urban Development.

27. Regional Plan Association. Human Responses to Patterns of Urban Density, New York, August 1969.

28. Sheldon, E. B., and W. E. Moore. Monitoring Social Change in American Society, pp. 10–12 in *Indicators of Social Change,* E. B. Sheldon and W. E. Moore (eds.). Russell Sage Foundation, New York, 1968.

29. Skinner, B. F. *Beyond Freedom and Dignity*, Alfred A. Knopf, Inc., New York, 1971.

30. Srole, L., et al. *Mental Health in the Metropolis: The Mid-Town Manhattan Study*, Vol. 1, McGraw Hill Book Co., New York, 1962.

31. Taeuber, C. Population: Trends and Characteristics, pp. 27-73 in *Indicators of Social Change*, E. B. Sheldon and W. E. Moore (eds.), Russell Sage Foundation, New York, 1968.

32. Taeuber, K. E., and A. F. Taeuber. *Negroes in Cities*, Aldine Publishing Co., Chicago, 1965.

ESTABLISHING PRIORITIES
AMONG ENVIRONMENTAL STRESSES

Howard Reiquam

Battelle Memorial Institute
Columbus, Ohio 43201

INTRODUCTION

All scientists are taught that to understand what they observe and then to predict or control the course of events, they must know the underlying causes. Whether it is for this reason, or simply because of a tendency to look for someone to blame for our problems, it is frequently stated that the root of the environmental difficulties we see everywhere about us is the Judeo-Christian anthropocentric ethic of man holding dominion over all other creatures and the earth itself. We are told that the American Indians and the Eastern religions hold nature in reverence and seek to minimize any conflict between man and the rest of the world. Yi-Fu Tuan[5] refutes this argument, pointing out for example that deforestation had become a serious problem in China well before the 3rd century B. C. In addition, it is known that North American Plains Indians sometimes fired the prairie in order to drive herds of bison over cutbanks where they could be butchered in quantity.

Evidently the ancient Chinese considered land clearing or timber utilization more important than soil conservation; similarly, the Blackfoot and the Sioux were presumably more intent on obtaining sufficient meat than on avoiding the risks of wild fires.

The point is, there seems to be little advantage in looking for some particular ethic or culture on which we can lay blame for the current mess we are making of the world. Perhaps we should simply accept that man is, by nature, oriented toward short-term gain. In any case, the solutions to these problems, that is, the control of this course of events, requires first the desire on the part of man to mend his ways. Increasingly, that desire is developing, but as more evidence of impending doom is made available, it becomes difficult to sort out that evidence and to know where to begin. Some disruption appears inevitable. We have been so wasteful of our resources — natural, human, and economic — that it is now necessary to assign some priorities to the problem in order to attempt to minimize the disruption.

Probably every individual has his own set of priorities with regard to the kinds of problems which are most important to him. In addition, within the particular kinds of problems of greatest individual interest, each person has some set of priorities again with which to rank the specific problems in the order that he chooses to worry about them.

Those of us who are especially interested in the environment tend to place it at or near the top of the first kind of list, but we have difficulty agreeing on a ranking scheme of the second kind, that is, among individual environmental problems. Most priority ranking schemes, in fact, have a certain narrowness of view reflecting the biases of the originators. This present effort is an attempt to combine some ideas others have had with some of my own in order to overcome part of that difficulty.

At risk of belaboring the obvious, it seems worthwhile to be explicit about what is meant by environmental problems in this context.

PROBLEM DEFINITION

The fact that politicians of virtually every persuasion are expressing interest in, or concern for, the environment is good evidence that interest at the public level is high. Beyond that expression of interest, however, there is a notable lack of agreement about the scale of the problem or even about a definition of the environment for which this common concern exists. Public appreciation of environmental problems is in some ways enhanced and at the same time made more difficult by statements and articles by various scientists who express their personal and sometimes conflicting views.

To some people, environmental degradation means little more than pollution of water and air. To others, the environment encompasses the entire biological, physical, and social milieu in which man is immersed, and thus, to enhance environmental quality is to give active consideration to all those features. The latter view is represented in the current wave of interest in ecology and ecological systems. Unfortunately, much of the alarm over alterations of the relationships between various organisms and their environment is based on incomplete understanding. Thus, in an effort to correct one imbalance, some other system is often stressed. To avoid that, there is a tendency to advocate the steady state solution in which a balance is struck among all interacting systems with no changes occurring. To achieve that steady state, however, the overall system must either be static or perfectly understood. In fact, of course, the total ecological system is poorly understood and is dynamic; so the real problems arise when the system is stressed to the point of upsetting some natural dynamic equilibrium. Moreover, different legitimate objectives, as Newbould points out,[1] represent different equilibria.

There are numerous viewpoints from which one can examine the question of environmental quality and ecological effects. For example, direct effects on human health are at the root of many of the standards which have been promulgated or proposed for various pollutants. Increasing efforts are being made to protect vegetation, fish, and wild life through stringent standards as well.

Another way of looking at the question, in hopes of gaining perspective, is to consider the scales on which effects appear. It is obvious that apparent effects are not necessarily (and in fact often are not) the most important. However, this approach can help illustrate some of the linkages that are of interest.

Easiest to note and consider are the localized pollution problems in which direct effects appear on a time scale short enough to permit ready association with sources. These are of course typified by visible wastes in surface water, smoke and dust in the air, and the like. Although secondary effects often are not significant locally, they can be in some cases. Photochemical smog is one example.

Regional scale problems usually include the direct effects observed locally, but because of dilution they are less evident. On this scale, the effects often shade over from acute to chronic, covering a wider area with reduced severity but often persisting longer. Secondary (and later) effects become more important, as, for example, the formation of photochemical smog, inadvertent meso-scale weather modification, altered population structures in streams, and eutrophication of lakes. Some of these effects are obvious and reasonably well understood; some are still little more than conjecture. Feedbacks appear in increasingly indirect ways in other subsystems of the environment.

Continental and global scale problems rarely appear as direct effects of specific acts or of pollutant emissions measurable at a particular point in space or time. Interplay of several subsystems is often involved, and the severity of specific effects is often open to question due to incomplete understanding; mechanisms are often unclear. The time scale stretches as effects work through various subsystems, sometimes with a multiplying or concentrating effect. Among the problems that have been suggested on this scale are worldwide climate modification, upset of evolutionary processes, or massive modification of the chemistry of surface waters, the atmosphere, and precipitation.

Although natural influences such as earthquakes and volcanic eruptions often have catastrophic environmental impact, attention is restricted here to those influences stemming from human activities which impinge on the environment.

Certainly it can be argued that priorities are most seriously lacking for an assessment of *activities* that are deleterious to the environment. Sooner or later, however, I believe that a more specific kind of examination is required to complete any assessment of activities. Probably no single approach is entirely satisfactory; there are many overlaps between individual environmental "stresses" and activities. In addition, there can be overlaps among stresses. For example, plastics in manufactured form can be non-degradable litter and solid waste or, when incinerated, can be the source of various toxic materials. In any case, there is room for argument over the best place to aggregate the components of the problem. An aggregated activity such as transportation, for example, could easily include such stresses as oil spills, a number of atmospheric pollutants, community noise, and several others. On the other hand, while an activity can be decomposed into individual stresses, they in turn can be decomposed into more and more specialized disciplinary problems. Traditional reductionists tend to favor this approach, breaking the extremely complex environmental system into a series of more or less elementary problems that lend themselves to a specialized approach.

However, there is an almost overpowering tendency to become so involved in the elementary aspects of the total problem that the links between elements, or the interactions among subsystems, with their feedbacks, are often overlooked or at least underestimated in their importance. Tschumi[4] attributes the widespread failure to foresee and take into account the full complexity of the problem to this traditional reductionist approach. He says that there is no doubt that the progressive destruction of natural ecosystems was not foreseen when man began to manipulate the environment and the growth rate of the population. He concludes that there is a scarcity of individuals who are familiar with the most important political, economic, social, and technological problem areas and who at the same time maintain an awareness of the biological implications.

The systems analysis approach to environmental problems, in which the total interlocking complex is considered, is sometimes suggested as the only way to arrive at balanced, comprehensive solutions. But a systems study requires essentially a decomposition of what is conceived to be the total system into a tractable set of inter-related and interacting subsystems. In the case of the environment, there is incomplete knowledge of the impact of subsystems upon each other; the linkages are poorly understood.

In a conscious effort to avoid aggregating too soon, and yet not get so interested in trees as to forget about the forest, the choice here has been to view individual environmental stresses relative to each other.

A FRAMEWORK FOR SETTING PRIORITIES

Probably all efforts in this direction suffer from the same difficulty in assigning relative numbers to essentially unquantifiable factors. One way to begin is to think in terms of graphs or networks first, and after being satisfied that the critically important relationships are somehow incorporated, then begin thinking of appropriate mathematical relations to describe them. Finally, we can think about algorithms which combine both mathematical and non-mathematical transformations. That sequential procedure seems to me to be more appropriate than to take the more usual mathematical approach to model design. Such a "morphological analysis" emphasizes fundamental structural differences and/or similarities rather than functional or performance features. In the beginning, then, empirical quantitative data on the past or current state of the factors in which we are interested are less important than structural data concerning individual mechanisms and sub-mechanisms.

Among the attempts at structuring a framework for comparing problems, no doubt the best known and most impressive is Platt's.[2] He classified a set of problems and crises according to an "estimated time to crisis" and an estimated "crisis intensity" which he found by multiplying some degree of effect by the approximate number of people affected by the problem. One of his crisis problems, which he expects to cause great destruction or change within the next 5 to 20 years, is "ecological balance," and a slightly less "intense" crisis he calls simply "pollution." The whole set of problems considered here falls within those two categories. (It should be pointed out that over two of those years have already passed since Platt wrote.) On a worldwide basis, he estimated that the crisis intensity is only one order of magnitude less for problems such as famine and ecological balance than for the total annihilation that would attend nuclear or RCB warfare escalation. Another order of magnitude lower in crisis intensity would be poverty and pollution.

For whatever reasons, Platt's action recommendations do not seem to have been adopted.

My own interests have prompted this attempt at decomposing ecological balance and pollution into individual environmental stresses. An idea central to this attempt is due to Royston.[3] He pointed out that if we view one problem, or stress, at a time, it amounts to isolating a specific stress S_i of the set $S_1 \ldots, S_n$. Stress S_i has an impact not only upon mankind, the resource pool, and the environment as a whole, but upon the rest of the stresses in the set as well. In turn, other members of the set have additional influence on stress S_i. The three-dimensional relationship with feedbacks is illustrated in Fig. 1.

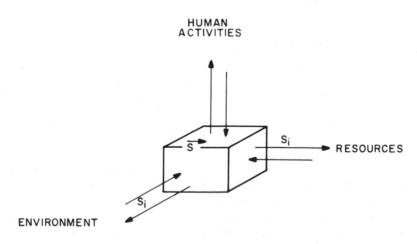

HUMAN
ACTIVITIES

Fig. 1. Interaction of human activities, resources, and the environment.

OPERATIONS WITHIN THE FRAMEWORK

The most difficult part of an attempt to relate the impact of a particular problem or stress upon mankind, resources, and the environment is to devise a common treatment or examination of those three different kinds of systems.

In the approach selected here, two somewhat subjective assessments are made for each environmental stress considered. They are common to the three major systems; a persistence index, P, ranging from 1 to 5, is assigned on the basis of physical and/or chemical characteristics. The reactive hydrocarbon component of automobile exhaust, for example, persists for only a few hours, while radionuclides such as carbon-14 have a half-life of thousands of years. A geographical index, R, between 1 and 5 is assigned according to the detectable range of the environmental stress in question as a function of the transportability and pervasiveness of sources of the stress. The persistence and range scales are shown in Table 1.

Table 1. Index scales for environmental assessment

Range index	Persistence index	Index value
Local	Days	1
Regional	Months	2
Continental	Years	3
Intercontinental	Decades	4
Global	Centuries	5

Table 2. Major systems and their components

Human	Environment	Resources
Biological	Land	Energy
Social	Air	Nutrition
Political	Water	Materials

A third index is assigned to describe the relative complexity of the impact of each stress upon each of the three major systems. Table 2 shows the components of each of the systems that are considered here.

The components of the human system relate to human reactions. Biological includes the physiological effects reflected in the health and physical well being of people. Social refers here to fears, alarms, general awareness, and "anti" activities. Political, in contrast, has reference to organized concern and group decision-making, in particular with respect to situations with international implications.

In the resource system, the components relate to availability. A given stress may affect the nutrition, raw materials, or energy pools. In addition, the energy component is considered for those stresses which stem from energy production.

We can consider land, air, and water as the environmental components, first as the media through which stresses are transmitted. Each component interacts with other components within its system and there is intersystem interaction as well.

The most obvious omission in this framework no doubt is an explicit consideration of economics. But economics pervades these systems; it .is therefore included implicitly.

A complexity index, C, is assigned to each of the systems according to how many of the components in that system are impinged upon in some way by stress S_i. In the present analysis, likely or expected as well as documented interactions are counted. These systems are extremely broad; so every one of the environmental stresses considered here for illustration interacts in some way with each of them. This index might be refined in various ways; for example, some group concensus (or an extremely selfconfident individual) might assign probabilities to the various interactions. In what follows, however, the indexes C, P, and R are discrete and strictly positive. That is, no distinction is drawn between "good" and "bad" interactions or effects.

Intuitively, it would seem that overall urgency or severity should be some function of "effects" integrated over time and space. However, one of the goals here is to help identify those problem areas where there is the greatest potential hazard and then to help identify and evaluate all of the effects. Therefore, the simple product of the three assessment scales can be used to indicate the relative importance of a given stress with respect to any one of the systems. That is, for example, the importance of stress S_i with respect to the environment, E, alone is

$$U_{i,E} = P_i R_i C_{i,E}$$

In this way, it is possible for something with relatively small effects, but spread over a very large area for a considerable period of time, to be as important or urgent as one which is highly localized but has implications in several of the components considered.

The R and P scales are quasi-logarithmic. Both indexes are dimensionless; they simply provide a consistent mechanism by which one can consider long- or short-range and immediate or long-term effects.

To obtain the C index for each major system, I have simply determined which of the broadly defined components within it interacts in some way under the influence of each environmental stress under consideration. For example, carcinogenic influences or lead poisoning represent biological interactions; corrosion of metals represents an effect on materials in the resource pool; effects on plant life are reflected in the nutrition component. Each effect can be related to one or more of the environmental components.

I have already mentioned the difficulty of assigning proper weights to the impact of a particular stress with respect to each of the human activity (H), resource (R), and environment (E) systems. The relative importance or weight to be assigned to $U_{i,E}$, $U_{i,R}$, and $U_{i,H}$ may always be a matter of personal values and convictions. At any rate, there does not appear to be an obvious way to eliminate value judgments completely.

My choice, in order to establish a common scale for overall ordering, is simply to assign equal weights to the three systems. That is, the overall complexity of a given stress is simply the sum of the system complexity indices.

$$C_i = C_{i,E} + C_{i,H} + C_{i,R} \, .$$

Therefore,

$$U_i = P_i R_i C_i \, .$$

RESULTS

Twenty widely recognized environmental stresses are used for illustration. Because the very serious potential damage over long periods of time on a global scale due to fallout alone from the use of nuclear weapons is considered to be totally unacceptable, it is included as a sort of benchmark. In Table 3, it is also used as a specific illustration of the procedure which has been used here for each stress.

Table 3. Interaction of stress components

Fission Products

$^{137}Cs, \; ^{90}Sr$

        LAND ← rainout, washout
    → AIR ← reentrainment
        WATER ← rainout, washout
        BIOLOGICAL ← mutagenic, carcinogenic effects
        SOCIAL ← fear, alarm
        POLITICAL ← intra- and international ramifications
        ENERGY
        NUTRITION ← effects on food chain
        MATERIALS

On the assumption that large quantities of fission products such as cesium-137 and strontium-90 would be released in a nuclear war, many half-lives would be required before decay to "acceptable" levels occurred. It is well known that weapons debris eventually becomes distributed globally; so both the range and persistence indices would be at the top of their respective scales. That is, P = 5 and R = 5. Seven of the nine components interact in some way, so C = 7. Thus, U = PRC = 175.

If, instead of being restricted to fallout, the stress had included the blast and thermal radiation effects of a nuclear exchange, then all of the components would have been included. In that case, C = 9 and U = 225, the maximum on all these scales. The intensity of effects would also be enormously greater, but in this scheme, the mere existence of an interaction or effect means that it gets counted.

Two priority lists of the other nineteen illustrative stresses are generated. One (Table 4) is based on my estimates of present practices in the developed countries, and the other (Table 5) is simply a guess based on estimates that include projected power demands and other increases in technological activities. Most of the changes from the list in Table 4 to that in Table 5 are due to increases in R, as the respective stresses become more pervasive. In several cases, C also increases because of the additional quantity or level which is expected.

Table 4. Environmental stress in descending
order of priority — "Present"

	P	R	ΣC	U
Pesticides	4	5	7	140
Heavy metals	5	2	9	90
Carbon dioxide	3	5	5	75
Sulfur dioxide (including oxidation products)	2	4	9	72
Suspended particulate matter	2	4	9	72
Oil spills	3	2	8	48
Waterborne industrial wastes	4	2	6	48
Solid waste	5	1	7	35
Chemical fertilizer	3	2	5	30
Organic sewage	2	2	6	24
Oxides of nitrogen	2	2	6	24
Radioactive waste (for storage)	5	1	4	20
Litter	4	1	4	16
Tritium, krypton-85 (nuclear power)	4	1	4	16
Photochemical oxidants	1	2	6	12
Hydrocarbons in air	1	2	5	10
Carbon monoxide	1	3	3	9
Waste heat	1	1	5	5
Community noise (including sonic boom)	1	1	4	4

Table 5. Environmental stresses in descending
order of priority – "Projected"

	P	R	ΣC	U
Heavy metals	5	3	9	135
Solid waste	5	3	8	120
Tritium, krypton-85 (nuclear power)	4	5	6	120
Suspended particulate matter	2	5	9	90
Waterbone industrial wastes	4	3	7	84
Carbon dioxide	3	5	5	75
Oil spills	3	3	8	72
Sulfur dioxide (including oxidation products)	2	4	9	72
Waste heat	3	4	6	72
Chemical fertilizer	3	3	7	63
Organic sewage	2	3	8	48
Oxides of nitrogen	2	3	7	42
Litter	4	2	5	40
Radioactive waste (for storage)	5	2	4	40
Pesticides	2	3	5	30
Hydrocarbons in air	1	3	6	18
Photochemical oxidants	1	3	6	18
Community noise (including sonic boom)	1	3	5	15
Carbon monoxide	1	3	4	12

Waste heat, from industrial processes and electric power plants, is notable in that one can reasonably expect it to move to a higher position on each index. Pesticides, on the other hand, will probably remain as a large-scale problem but should fall down the list simply because the use of persistent pesticides will be restricted. In addition, however, their toxic specificity should be improved.

SUMMARY

No attempt has been made here to rank the seriousness of specific effects such as chronic lead poisoning or emphysema. Neither are "good" and "bad" effects or interactions considered. That is deliberate, in the attempt to avoid imposing a particular value system on the ranking scheme.

In most cases, more questions than answers are suggested by the analysis described here; the ordering is intended to help identify and put into perspective those problems with the widest and most complex implications or manifestations. The logical next steps are to attempt a systematic assessment of effects; to identify the biological, physical, and social manifestations of each stress; and then to mitigate those that are less than beneficial and at the same time enhance the favorable ones. That, obviously, is what many individuals and organizations are attempting to do now, but often those efforts appear to be narrowly defined, with little or no attention being paid to some of the potentially serious long-range, long-term problems.

Perhaps the same framework used here for setting priorities can be used to help structure the broad, interdisciplinary effort that is needed. A loosely structured group of individuals with broad and comprehensive understanding of the environment and of the biological, physical, social, and engineering sciences (as suggested by the components listed in Table 1) should be able to examine activities that impinge on the environment through particular stresses, identify most of the interactions which occur, and probably most important, identify the major gaps in our knowledge in order to recommend specific problems for detailed study and basic specialized research. If it is true that some disruption is inevitable as the pervasive environmental problems are attacked, then a holistic approach as suggested here should suggest tradeoff criteria which might help minimize the disruption.

ACKNOWLEDGEMENT

Most of the thinking which went into the development of these ideas was done with a Visiting Fellowship at Battelle Seattle Research Center. That support is gratefully acknowledged.

REFERENCES

1. P. Newbould, How to Manage an Ecosystem, *New Sci.*, 47: 191–193 (July 23, 1970).
2. J. Platt, What We Must Do, *Science*, 166: 1115–1121 (Nov. 28, 1969).
3. M. G. Royston, Institut Battelle, Geneva, personal communication, June 1971.
4. P. A. Tschumi, Will the Exploding Human Population Succeed in Conserving Nature?, *Experientia*, 26: 572–576 (1970).
5. Yi-Fu Tuan, Our Treatment of the Environment in Ideal and Actuality, *Amer. Sci.*, 58: 244–249 (1970).

BIBLIOGRAPHY

Ayres, R. U. *Technological Forecasting and Long-Range Planning*, McGraw-Hill, New York, 1969.
Bauer, R. A. *Second-Order Consequences*, MIT Press, Cambridge, 1969.
Bereano, P. L. The Scientific Community and the Crisis of Belief, *American Scientist*, 57: 484–501 (1969).
Brooks, N. H. Man, Water, and Waste, pp. 91–112 in *The Next Ninety Years*, California Institute of Technology, Pasadena, 1967.

Chapman, J. D. Interactions Between Man and His Resources, pp. 31–42 in *Resources and Man*, by the Committee on Resources and Man, National Academy of Sciences–National Research Council, Freeman, San Francisco, 1969.

Coale, A. J. Man and His Environment, *Science*, 170: 132–136 (October 9, 1970).

Crowe, B. L. The Tragedy of the Commons Revisited, *Science*, 166: 1103–1107 (November 28, 1969).

Dubos, R. *So Human an Animal*, Scribners, New York 1968.

Dubos, R. *Reason Awake: Science for Man*, Columbia University Press, New York, 1970.

Glass, B. Biological Aspects of Technology Assessment, *Quart. Rev. of Biology*, 45: 168–172 (1970).

Hardin, G. The Tragedy of the Commons, *Science*, 162: 1243–1248 (December 13, 1968).

Mayr, E. Biological Man and the Year 2000, pp. 200–204 in *Toward the Year 2000*, D. Bell (ed.), Houghton, Mifflin, Boston, 1968.

Moncrief, L. W. The Cultural Basis for our Environmental Crisis, *Science*, 170:508–512 (October 30, 1970).

Rabinowitch, E. The Mounting Tide of Unreason, *Bull. Atomic Scientists*, 4–9, May, 1971.

Lord Ritchie-Calder. Mortgaging the Old Homestead, *Foreign Affairs*, 48: 207–220 (1970).

Russell, C. S., and H. H. Landsberg International Environmental Problems – a Taxomony, *Science*, 172: 1307–1314 (June 25, 1971).

Spofford, W. O., Jr. Closing the Gap in Waste Management, *Environmental Science and Technology*, 4: 1108–1114 (1970).

Starr, C. Social Benefit vs. Technological Risk, pp. 24–39 in *Proceedings of Symposium on Human Ecology*, U.S. Dept. of Health, Education, and Welfare, Washington, D. C. 1968.

Wagar, J. A. Growth vs. the Quality of Life, *Science*, 168: 1179–1184 (June 5, 1970).

White, L., Jr. The Historical Roots of our Ecological Crisis, *Science*, 155: 1203–1207 (March 10, 1967).

The Unforeseen International Ecologic Boomerang, Special Supplement, M. Taghi Farvar (ed.). *Natural History*, pp. 41–72, February 1969.

POLLUTANT BURDENS IN HUMANS:
A MEASURE OF ENVIRONMENTAL QUALITY

John H. Finklea

Director, Division of Health Effects Research
National Environmental Research Center
Environmental Protection Agency
Research Triangle Park, North Carolina 27711

Douglas I. Hammer

Chief, Epidemiology Branch
Division of Health Effects Research
National Environmental Research Center
Environmental Protection Agency
Research Triangle Park, North Carolina 27711

Kenneth M. Bridbord

Epidemiologist, Division of Processes and Effects Research
Office of Research and Monitoring
Environmental Protection Agency
Washington, D. C. 20460

Vaun A. Newill

Special Assistant for Health
Office of Research and Monitoring
Environmental Protection Agency
Washington, D. C. 20460

INTRODUCTION

Pollutant burdens in humans are becoming important milestones in our effort to assure that the health of our citizens and the quality of our environment will be protected from the adverse effects attributable to an increasingly urbanized and industrialized society. Human pollutant burdens will find greatest utility as tools for the appraisal of multimedia environmental problems related to metals, pesticides, and synthetic organic compounds.[1] By definition, a tissue carries a pollutant burden whenever it contains an environmental residue greater than that needed for optimum growth and development. Each of us has multiple pollutant burdens. Human pollutant burden patterns may serve as indicators of the environmental quality, as

83

reflectors of biological response, as inputs into environmental standards, as channel markers for research, and as safeguards for pollution control and recycling technology.

AS ENVIRONMENTAL MONITORS

Human tissues are dose integrators for environmental pollutants. Everyone maintains a series of involuntary personal dosimeters recording pollutant exposures for different time intervals. Those who would utilize these dosimeters must appreciate the underlying, interlinked exposure and metabolic factors that determine tissue burdens.[8] For example, metal exposures usually involve a number of different chemical compounds impinging upon man through multiple environmental media. Our intake of cadmium involves air, water, food, and tobacco smoke. Pollutant absorption is likewise a function of exposure route, physical form, and chemical composition. For example, lead fume is rapidly and almost completely absorbed from the lung and lead bound to respirable particulates is largely absorbed. Conversely, lead bound to large suspended particulates fails to penetrate deeply into the respiratory tract and is shunted by the mucociliary apparatus to the gastrointestinal tract where absorption is relatively limited. In addition, organic lead compounds are much more readily absorbed and more toxic than inorganic lead compounds.[3]

Distribution, storage, and excretion kinetics are generally more complicated than those of exposure and absorption. Biological half-lives of pollutants greatly differ; for mercury it is 80 days, for lead 10 years, and for polychlorinated biphenyls probably longer than a lifetime. Acute pollutant exposure and mobilization of residues from depots are usually measured by blood and excreta levels. Even then, different metabolites may indicate different processes. Plasma DDD and urinary DDA are better indicators of recent exposure than plasma DDE which may be of dietary origin or recently mobilized fat. Blood is a transport tissue that assumes less importance when one is concerned about chronic low-level exposures, increasing total body burdens, and subtle adverse biological effects. Depot tissue may also record isolated acute exposures as shown by metaphyseal lead lines and hair arsenic bands.[13,18]

Different tissues selectively concentrate different pollutant residues. Tissues with high lipid content like fat and brain serve as depots for ^{85}Kr, alkyl mercury, chlorinated hydrocarbon pesticide, and polychlorinated biphenyl plasticizer residues. The liver and kidney retain much of the body cadmium burden, while lead, selenium and strontium concentrate in bones and teeth. Similarly, the lung retains a number of pollutants, including asbestos and chromium. Within a tissue, the location of a pollutant is undoubtedly of critical importance. For example, pollutants bound to cell membranes would be expected to have different effects than those bound to intracellular organelles. Monitoring pollutant burdens of intracellular structures in humans is not now a practical indicator of environmental quality. When sampling living populations, the epidemiologist must focus on more available specimens including hair, blood, excreta, and placenta.

Pollutant burden appraisals have been applied to the metallic pollutants arsenic, cadmium, copper, lead, manganese, mercury, strontium, and zinc.[2,6,14-17] They should be equally useful for antimony, beryllium, chromium, nickel, selenium, silver, titanium, tin, and vanadium. Human pollutant burden patterns have also been sketched for the persistent pesticides DDT, dieldrin, lindane, and heptachlor.[5] Pollutant burdens related to other synthetic organic compounds — such as colorants, flavors, plastic products, rubber products, surfactants, and organic intermediaries — have received insufficient attention.

Of prime current concern are polychlorinated biphenyl (PCB) compounds, phthalates, chlorinated naphthalenes, chlorinated aliphatics, and hexachloro-benzene.[4] Not all pollutants can be approached through pollutant burden studies. For example, organophosphate pesticide exposures and possibly exposures to certain fuel additives can best be assayed by changes in cholinesterase levels even though sensitive methods are now being developed to measure their polar metabolites in urine. Residues of optical brighteners, important constituents of soap and detergent products, would likewise be difficult to determine in human tissue or excreta. Similarly, pollutant burden studies would not seem valid for oxidant air pollutants or sulfur dioxide. Carbon monoxide, nitrogen dioxide, and suspended particulate air pollution occupy an intermediate position. Recent but not remote exposure of non-smokers to carbon monoxide and nitrogen dioxide might be assayed by carboxyhemoglobin and methemoglobin levels, and particulates in lung have been quantitated at necropsy.

On occasion human tissues, serendipitously preserved, have furnished important insights into current pollution problems.[10,11] Environmentalists have utilized museum specimens to elucidate pesticide, plasticizer, and trace element pollution problems. There is a pressing need for human tissue banks to build an environmental flashback capability and to serve as an environmental alarm system. There are several thousand metallic and synthetic compounds that potentially involve human exposures and human pollutant burdens. Infectious disease investigators encountered similar problems with unknown agents three decades ago and set up a network of sera banks that permitted them to assess rapidly the relationship of newly discovered microbes to older microbes, to disease syndromes, and to the factors of age, sex, race, residence, and time. A coordinated tissue bank program could overcome the obstacles inherent in standard tissue preservation. Freezing, lyophilization, and irradiation seem the methods of choice, with special care being devoted to containers. Less dramatic but most useful tissue banks would involve easily preserved tissues like hair, deciduous teeth, and nail clippings collected from living populations whose environmental exposures are concomitantly measured.

AS INDICATORS OF BIOLOGICAL RESPONSE

Human pollutant burdens should be related to biological responses, which may be placed in perspective by several bench marks. In general, one thinks of five stages, as shown in Fig. 1: (1) a pollutant burden not associated with other changes, (2) a pollutant burden associated with physiologic changes of uncertain significance, (3) a pollutant burden associated with physiologic changes that are disease sentinels, (4) a pollutant burden associated with morbidity, and (5) a pollutant burden associated with mortality. Even where pollutant burdens have been measured, only fragmentary information is available to link most burdens with biological responses. Our laboratory colleagues have rarely focused their attention upon biological response methods that can be applied to living populations. A concerted effort should be mounted to discern sensitive biochemical and physiologic indicators that may then be related to pollutant burdens. Current studies of the effect of cadmium upon the zinc metalloenzymes alkaline phosphatase, carbonic anhydrase, leucine amino-peptidase, and carboxypeptidase are a useful beginning.

The five-stage biological response spectrum is a useful conceptual framework. There are, however, limitations other than information gaps. Not all disease processes fit comfortably into the response gradient, as illustrated by teratogenesis. Pollutant imprinting may also occur. Here the latency period for the response is longer than the biological life of the pollutant residue. In

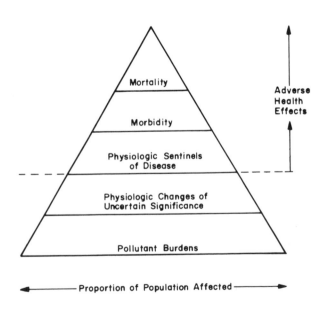

Fig. 1. Spectrum of biological response to pollutant exposure.

neurologic disorders associated with manganese poisoning, all tissue manganese levels are not elevated. Similarly, exposure of neonatal animals to a variety of short-lived carcinogens may initiate late neoplastic changes.

Despite the many underlying complexities, useful deductions may be made from pollutant burden patterns. Prudence is mandatory when interpreting the general population patterns. A given pollutant burden might represent a minor risk for an industrial population of medically prescreened adults and a major risk for susceptible groups in the general population. Susceptible groups include the pregnant mother and her fetus, infants and children, the elderly, chronic disease cases, and patients with genetic diseases such as deficiencies of glucose-6-phosphate dehydrogenase and alpha$_1$ antitrypsin.

Age patterns can help identify special risk problems. For example, PCB residue levels rise with age before dropping precipitously during the eighth decade of life.[4] A superficial, unjustified interpretation might be that survivors have low levels and that the residue may be linked with a lethal disorder. In this case, the oldest sample was quite small, and thus their mean PCB residue level was not reliable. This study also illustrates the need to understand residue metabolism. PCB residues are most likely bound by plasma lipids and thus may be innocently associated with important atherosclerosis risk factors. The PCB data represent a cross-sectional view of the time factor. Another method, the cohort approach, involves repeated sampling of the same population over time. Cross-sectional studies actually include several age cohorts.

Residues may have different sex patterns as shown by plasma DDT levels where residues in males exceeded those observed in females at every age. This male excess may have been due to greater exposure or differing metabolism. Ethnic group and residence may likewise effect pollutant burden patterns. Black children had total plasma DDT residues greatly exceeding those of white children. Rural black children actually equalled levels found in pesticide formulators who are the most heavily exposed occupational group. Clearly, these children are more likely to suffer any ill effects related to DDT residues. On the other hand, rural blacks exhibit plasma PCB residues less frequently than urban whites.

Differences between cities are also found, as illustrated by the study of environmental exposure and hair trace-element levels.[8] The non-essential trace metals arsenic, cadmium, and lead followed an environmental exposure gradient while the essential trace metals copper and zinc, for which homeostatic mechanisms are well developed, did not significantly vary along a more limited exposure gradient. Pollutant burden surveys may also reveal individuals with residue patterns that warrant special investigation to abort clinical disease and to determine unusual environmental exposure routes.

AS INPUTS FOR STANDARDS DEVELOPMENT AND APPRAISAL

One method of environmental protection will be the promulgation and enforcement of ambient and emissions standards. Standards for pollutants characterized by direct exposure from a single environmental medium are set by considering control technology and the associations between pollutant levels and adverse effects and then by allowing a safety factor. Sets of standards for pollutants impinging upon man through multiple environmental media are much more difficult to derive. One input into such standards will be the pollutant burden pattern of populations at greatest risk. For these pollutants, routine monitoring of population burdens may prove at least as important as environmental monitoring. Population burdens may also delineate a priority order for standards development.

Critical evaluation of each set of standards will be necessary, and changes in human pollutant burdens will be one way to discern whether standards actually are achieving environmental quality goals. Increasing pollutant burdens would warn that environmental controls are inadequate. Should population burdens approach levels associated with clinical toxicity, environmental standards would have failed and emergency action would be indicated.

AS CHANNEL MARKERS FOR RESEARCH

Pollutant burden patterns can serve as channel markers for research. Ubiquitous pollutants, even those that appear benign in acute toxicity studies, might increase the risk or hasten the onset of chronic diseases. Demonstrable pollutant burdens in newborns and young children should alert investigators to teratogenic, carcinogenic, and mutagenic hazards. Peak levels at puberty should signal concern for reproductive performance. Laboratory evaluation of these hazards will be difficult and expensive. Selection of chemicals for toxicologic study will be critically important. Pollutant burden studies should contribute to the selection process. Conversely, laboratory studies and environmental assays can help select which pollutant burdens should be studied and with what frequency.

AS SAFEGUARDS FOR ENVIRONMENTAL CONTROL AND RECYCLING TECHNOLOGY

Recently advocated pollution control techniques, especially those involving reuse and recycling, may result in an increased exposure to multimedia toxic substances. Alkyl manganese compounds have been advocated as a substitute for lead in fuel additives; waste paper containing appreciable quantities of selenium has been suggested as a fuel for power plants; newer incinerators may well convert and emit polychlorinated biphenyls as respirable particulates; and incineration of sewage sludge to avoid ocean dumping can increase many atmospheric trace element levels. In each of these cases, no adequate model involving exposure and biological response exists to allow a quantitative appraisal of the potential effects upon human health. Expanded human pollutant burden studies coupled with the resources of a national tissue bank are crucial inputs for such models. The feasibility of recycling treated sewage water directly into municipal reservoirs and consuming food produced from solid wastes has been questioned by those who fear technology will not sufficiently safeguard the consumer. When recycling systems are utilized, appropriate human pollutant burdens should be monitored along with the recycled products to confirm that human health is actually protected.

AN OVERVIEW OF RESEARCH

The Environmental Protection Agency (EPA) currently mounts human pollutant burden studies of the two types. One series of investigations is designed to determine which tissues are the most suitable indicators of environmental quality. Here blood, urine, hair, excreta, and placenta are assayed for selected trace metals or synthetic organic compounds. In schoolboys, hair but not urine reflected community exposures to arsenic, cadmium, and lead. Incidentally, house-dust levels of pesticides and trace metals are good predictors of blood and hair levels. Conversely, blood but not hair proved a useful indicator of PCB exposure in junkyard workers. Maternal—fetal tissue sets including placenta, mother's blood, and cord blood from the newborn baby are useful monitors for lead, cadmium, mercury, and chlorinated hydrocarbon pesticides. Current investigations involve other pollutant residues in the same tissues. Multiple tissue sets are also being collected at necropsy. Whenever possible, study groups are chosen because of their exposure to an appropriate pollutant gradient. The second series of investigations tests specific hypotheses. Maternal blood and placental cadmium levels were not found increased in toxemia of pregnancy. Renal cadmium levels were not consistently elevated in cardiovascular disease decedents. Previously observed apparent elevations in renal cadmium may have been occasioned by the confounding effects of the widely varying renal cadmium levels noted in cancer deaths.[7,9] Blood pressure and serum lipids were inconclusively related to occupational and domestic cadmium exposure.[7] Current investigations test the relationship of PCB and trace metal residues in surgical biopsies and necropsy sets to air, water, and industrial exposures with appropriate consideration of personal covariates and medical diagnosis. Placental and hair levels of trace metals collected from volunteers residing in 30 urban communities are being related to house-dust, air and water levels of these pollutants. Attempts are being made to link pollutant burdens with refined laboratory indices of health impairment. One promising index is a placental enzyme activity profile which is being structured with two distinct components. One component consists of enzymes induced or altered by specific pollutants, while the other includes enzymes that are rate limiting for selected metabolic cycles. Changes in the latter component could serve as a sentinel for occult environmental problems.

Future refined indices might include changes in serum metalloenzyme kinetics, induction of transport proteins, minor aberrations in lymphocyte karyotypes, changes in reflex patterns, and alteration of performance or behavior.

SUMMARY

Pollutant burdens in human tissues are crucial indicators of environmental quality. Tissue burdens integrate environmental exposures from multiple routes, providing an excellent assay system for many environmental trace metals and synthetic organic residues. Pollutant burdens link routine environmental monitoring to the risk of adverse health effects. Pollutant burden assays, coupled with a national tissue bank, can provide important intelligence for the setting and appraisal of environmental standards. Environmental control procedures, including reuse and recycling of waste, should be evaluated by surveillance of appropriate human pollutant burdens. The current relevant Environmental Protection Agency research program is reviewed.

REFERENCES

1. H. R. 5390. Toxic Substances Control Act of 1971, March 2, 1971.
2. Bidstrup, P. L. *Toxicity of Mercury and Its Compounds.* Elsevier Publishing Co., New York, 1964.
3. Engel, R. E., D. I. Hammer, R. J. M. Horton, N. M. Lane, and L. A. Plumlee. Environmental Lead and Public Health. EPA, APCO, AP No—90, March 1971.
4. Finklea, J. F., J. P. Creason, L. E. Priester, T. Hauser, and T. Hinners. Polychlorinated biphenyl residues in human plasma exposure – a major urban pollution problem?, abstract submitted for 99th Annual APHA Meeting, October 11–15, 1971.
5. Finklea, J. F., L. E. Priester, S. H. Sandifer, and J. E. Keil. South Carolina Community Pesticides Study Report, unpublished, 1969.
6. Gaffney, G. W., et al. Strontium-90 in Human Bone from Infancy to Adulthood, 1962–1963, *Radiol. Health Data Reports* 7: 383–386 (1966).
7. Hammer, D. I., J. F. Finklea, J. P. Creason, S. H. Sandifer, J. E. Keil, L. E. Priester, and J. F. Stara. Cadmium Exposure and Human Health Effects, in *Trace Substances in Environmental Health – V.* D. D. Hemphill, (ed.) University of Missouri Press (in press).
8. Hammer, D. I., J. F. Finklea, R. H. Hendricks, C. M. Shy, and R. J. M. Horton. Hair Trace Metal Levels and Environmental Exposure, *Amer. J. Epid.* 93(2): 84–92 (1971).
9. Hammer, D. I., A. W. Voors, K. Bridbord, M. Schuman, C. Pinkerton, and V. Hasselblad. Tissue Trace Metals and Cardiovascular Disease: Cause or Consequence?, in preparation.
10. Jaworowski, Z. Stable Lead in Fossil Ice and Bones, *Nature* 217: 152–153 (1968).
11. Jensen, S. PCB as Contaminant of the Environment – History, PCB Conference, September 29, 1970, National Swedish Environment Protection Board, Research Secretariat, Solna, December 1970.

12. Morgan, D. P., and C. C. Roan. Chlorinated Hydrocarbon Pesticide Residue in Human Tissues, *Arch. Environ. Health* 20(4): 452–457 (1970).
13. Nelson, W. E. (ed.). *Textbook of Pediatrics.* W. B. Saunders Co., Philadelphia, 1969.
14. Schroeder, H. A., and J. J. Balassa. Abnormal Trace Metals in Man: Cadmium, *J. Chron. Dis.* 14: 236–258 (1961).
15. Schroeder, H. A., and I. H. Tipton. The Human Body Burden of Lead, *Arch. Environ. Health.* 17: 965–78 (1968).
16. Underwood, E. J. *Trace Elements in Human and Animal Nutrition*, Chapter 3, Copper; Chapter 6, Zinc; Chapter 7, Manganese, pp. 48–99, 157–186, 187–217. Academic Press, New York, 1962.
17. Vallee, B. L., D. D. Ulmer, and W. E. C. Wacker. Arsenic Toxicology and Biochemistry, *Arch. Industr. Health* 21: 132–151 (1960).
18. Wintrobe, M. M., et al. (eds.). *Harrison's Principles of Internal Medicine.* McGraw Hill Book Co., New York, 1970.

AQUATIC COMMUNITIES AS INDICES OF POLLUTION

Ruth Patrick

Chairman, Department of Limnology
Academy of Natural Sciences
Philadelphia, Pennsylvania 19103

Aquatic communities are very similar to terrestrial communities in many of their characteristics. For example, they are composed of organisms that decompose detritus that enters the system; primary producers that fix the sun's energy; herbivores — animals that feed upon the primary producers; carnivores, which feed upon herbivores; and omnivores which may feed upon carnivores, herbivores, and primary producers.

The energy efficiency of this food web is about the same as that for terrestrial communities, that is, about 10 to 15% of the energy fixed as protoplasm in primary producers is passed on to the herbivores and 10% from herbivores to carnivores. In some estuaries, the primary producers are more efficient than they are in terrestrial communities or some aquatic communities. Thus, this food web consists of four to five stages of energy transfer.

In each stage of energy transfer, we find many species belonging to many different systematic groups. This is true of the detritus feeders which may be bacteria, fungi, or various kinds of invertebrates. Likewise, it is true of the algae which may belong to the green algae, blue-green algae, diatoms, dinoflagellates, etc.; and it is also true of the herbivore and carnivore levels which may be protozoans, various types of worms such as flatworms and oligochaetes, various members of the mollusca and crustacea groups, and of course many insects and fish.

When we examine each of these stages of energy transfer, we find organisms that have various life spans; some of them reproduce one every day or every few days, whereas others may take months or even a year or two to complete their life cycle. Thus, ecological requirements are quite different, and it is this variability that brings about changes of species with the seasons of the year or with other types of environmental change. If we follow these species through their life histories, we find that the stage in their lives when they take most of their nutrient from the environment also varies and that at different stages they may be herbivores, carnivores, or omnivores. It is perhaps this great variability in life history, in ecological requirements, and in preferences for various types of prey which gives stability to the system; for there are many energy pathways and each of them may be affected by varying environmental conditions and predator pressures. This high variability also allows species always to be present and interacting in the ecosystem. We have found in our many studies of rivers that the numbers of species performing these various functions stay fairly similar throughout the year and from year to year in a given area, whereas the kinds of species vary greatly.

Typically in natural streams, we found that the populations of species are large enough to take care of the variations of the environment and of predator pressure which may occur and yet insure the reproduction of the species. Rarely does one find in truly natural streams a large population of any one species, and if such does occur, as in the algal blooms in the spring of the year, they are usually quickly reduced by predator pressure. It is only in polluted conditions where predators have been removed that one will find the same species in great abundance lasting for many months of the year. An example of this is *Stigeoclonium lubricum*, which is often found in polluted streams and can be found at almost any season of the year under such conditions.

Of the various species composing these ecosystems, the ones that most clearly reflect the chemical changes in the water are the primary producers, because they use the minerals in the water as a source of nutrition. The herbivores and carnivores more often reflect changes in physical conditions such as changes in current, temperature, and dissolved oxygen.

EFFECTS OF POLLUTION

Various types of pollution affect these aquatic communities in various ways. Examples of the types of change that one might expect can be seen from our studies and those of Tarzwell and Gaufin,[15] Ellis,[7] and many others. We have found in streams in the eastern United States that the first effect of an increase in a mineralized organic load — that is, the first increase in various chemical radicals such as nitrates, phosphates, ammonia, various trace metals, etc. — is to cause certain species to become more common. For example, in the spring of the year, such diatom species as *Navicula cryptocephala* and *Cyclotella meneghiniana* may become fairly common, whereas in the summer and early fall of the year *Melosira varians* may become very common. The protozoa change from a well balanced representation of amoebas, flagellates, and ciliates to a representation in which the ciliates are more common. The herbivore and carnivore populations are mainly represented by larger populations, and there is not very much of a shift in the kinds of species. Thus, one typically finds a wide assortment of species representing many different phylogenetic groups. Usually in the eastern United States, the insects — so far as numbers of species are concerned — dominate the invertebrate fauna except for the protozoa which may also be represented by many species. The fish in such streams, if they are warm water streams, represent a wide variety of types with largemouth bass, smallmouth bass, various types of sunfish, minnows, suckers, catfish, etc. being present. If the organic load is heavier, we see a decided shift from a diatom-dominated flora to one that has large populations of green algae, particularly species belonging to the genera *Oedogonium* and *Cladophora*. Under these conditions, we often find large beds of rooted aquatics such as *Anacharis* (Elodea), *Ceratophyllum*, and various types of *Potamogeton*. These beds of rooted aquatics will not be present unless the river is fairly clear so that the photosynthetic zone extends to the littoral zone. If this heavy organic load is mostly mineralized and contains no toxic pollution, the invertebrates may have large populations of *Physa* snails, Sphaerid clams, *Planaria*, crayfish, hellgrammites, chironomids, and blackfly larvae. Of course, not all of these may be present. What species are present depends upon various ecological conditions, but usually under such conditions, one or more of these groups are common. On the other hand, mayflies and stoneflies are usually represented by a relatively few species and often may be absent. The most common species of fish are carp and catfish, though bluegills may be fairly common. Typically one does not find as many species of fish under these conditions; however, those that are present may be represented by large populations. Under very heavy organic pollution, wherein the organic load has not mineralized but rather is in the state of decomposition grading into the mineralized zone, we typically find many thick beds of blue-green algae, particularly *Schizothrix calcicola* and *Microcoleus vaginatus*. Also, one may find plankton populations of *Anacystis cyanae* and *Anabaena aeriginosus*. The last two species are often found in lakes; however, in the Savannah River and also in the Potomac, in areas of heavy organic pollution, these species are very common in late summer. Under such conditions — that is, in the zone merging from degradation to mineralization — one typically finds *Sphaerotilus* if there are large amounts of carbohydrates in the organic load. In experiments carried out at the Stroud Water Research Center, we have found that concentrations of .05 to .1 mg/liter of glucose will bring about extremely heavy growths of *Sphaerotilus* in natural flowing streams. Under such conditions, the protozoan fauna is usually dominated by ciliates and *Carchesia* may be common. Rattail maggots and blackfly larvae may be very common, and in the sediments one often finds fairly large populations of *Chrionomus plumosa* and tubificid worms. These latter two groups are often found in areas with low dissolved oxygen but are not found if toxic pollution is present.

Beeton[1] and Davis[6] have summarized the types of changes that occurred in Lake Erie west of a line drawn between Point Pelee, Ontario, and Sandusky, Ohio. Since 1900, the population and industrial development in the watershed of western Lake Erie has expanded rapidly, and the lake receives large quantities of sanitary, industrial, and agricultural organic wastes. The lake has become modified by increases in dissolved solids and decreases in transparency and dissolved oxygen, Blooms of blue-green algae[6] of varying size increased markedly in the 1960's. Between 1919 and 1963, the abundance of phytoplankton increased threefold; spring and fall maxima were greater and lasted longer, and different diatom genera became dominant. The summer minimum became shorter and the winter minimum failed to materialize in later years. The spring pulse in 1920 was dominated by *Asterionella*, and this for the most part continued until 1956, when the bloom was dominated by *Melosira*.

Since that time (through 1963), *Melosira* has been the dominant or shared the dominance with *Fragilaria* (1958), except for 1960 when *Tabellaria* and *Melosira* were dominant. The fall pulse was largely *Synedra* or *Melosira* or a combination of the two from 1920 to 1944. In 1947 and in 1949, *Pediastrum*, probably *boryanum* or *simplex*, which are found in waters with a higher nutrient load, was a co-dominant. In 1958 and 1961, *Anabaena* was a co-dominant, and in 1962, *Anabaena* and *Oscillatoria*, which indicate nutrient-rich waters, were co-dominant. In 1963, the most important algae in the fall pulse were *Fragilaria*, *Melosira,* and *Stephanodiscus.* During the summer minimum, there has been an increase in *Pediastrum* and *Anabaena.* Copepods and Cladocera increased in abundance between 1939 and 1958. *Diaptomus siciloides*, which is characteristic of nutrient-rich water, has become a dominant in the zooplankton.

The shift in benthic invertebrates correlated with increased nutrients in western Lake Erie has resulted in the almost complete disappearance of the burrowing mayfly since 1930 and a great increase of oligochaete worms (*Aulodrilus* spp., *Brachiura sowerbyi, Limnodrilus hoffmeisteri, Peloscolex ferox*, and *Polamothrix* sp.) Pollution-tolerant species of fingernail clams, midges, and snails were more abundant.[5] Brown,[2] using the number of oligochaetes per m^2, concluded that between 1930 and 1951 the zone of heavy and moderate pollution had moved 5.5 miles and 8 miles, respectively, lakeward from the Maumee River.

Although Lake Erie continues to produce about 50 million lb of fish annually, there has been a great shift in kinds of species caught. In 1899, the catch was largely lake herring (Cisco), blue pike, carp, yellow perch, sauger, whitefish, and walleye. During the past 25 years, blue pike, lake herring, sauger, and whitefish have almost disappeared. In 1965, the fisheries in order of importance consisted of yellow perch, smelt, sheepshead, white bass, carp, catfish, and walleye. As often occurs, the shift has been in relative importance of species with the ones more tolerant of organic pollution and resultant changes in the food web out-competing the less tolerant species.

Toxic pollution is of many different kinds and occurs in varying quantities. Thus, it is hard to predict what the kinds of changes will be when toxic pollution is present unless one defines the type of toxic pollution. However, there are certain general trends which occur in the presence of toxic pollution. The reproduction of species may be inhibited but the species may not be killed immediately, although, of course, eventually the species is eliminated. Examples of this effect were shown in experiments[13] with diatom communities in which the pH of the circumneutral stream was lowered from 7 to 5.5. The various diatom species for the most part were not killed, but they ceased to reproduce. Thus, a relatively high number of species with small populations were present, yielding a fairly high diversity index; but the biomass was extremely low, and of course with time the community was eliminated.

Toxicants such as insecticides in very small quantities have been shown to affect the physiology and behavior of fish. For example, DDT is known to affect the nervous system of fish.[17] Cairns and Scheier[4] have shown that dieldrin affects the sight of fish. Such toxicants reduce the ability of a species to compete in the natural environment and thus bring about species change in the structure of the community. More toxic pollution usually eliminates many species, and the few that can survive develop fairly large populations. For example, in the Clinch River, when high conductivity resulted from the release of salt into the river, an overabundance of *Cladophora* developed and fairly large populations of dragonflies were found. Many species of invertebrates, including large populations of unionid clams, were eliminated. In Lititz Run, where the presence of small amounts of cyanide brought about the death of practically all aquatic organisms, *Stigeoclonium lubricum* continued to live and produce fairly large populations. Likewise in streams such as some of the headwaters of the Schuylkill River, which are affected by large amounts of acid mine wastes, we have found that practically all the herbivores and carnivores were eliminated. Only very rarely did we find any animals present in the stream; however, the algae *Stigeoclonium lubricum* and a species of *Tribonema* were common on rocks. Under such conditions, if sufficient nutrients are present, large beds of *Potomageton confervoides* may develop. This species typically lives in acid waters.

The effects of temperature upon aquatic communities are variable. In studies carried out in the eastern part of the United States, we have made the following kinds of observations. Small increases in temperature often bring about larger populations of the species that are present, and thus the total biomass of the community increases. An example of this type of change has recently been pointed out by Patrick[10] in which she has shown that diatom communities of White Clay Creek had a wide range of temperature tolerance, within which there was an optimum range. So long as the changes in temperature were not very large but approached the optimum, there was no deleterious effect upon the diatom community typical of White Clay Creek, as the numbers of species and population sizes increased slightly. However, when the temperature exceeded the optimum and rose toward the upper limits of the tolerance of the community, there was a decrease in species and a decrease in biomass. In other studies carried out by Patrick et al.[11] in White Clay Creek, we have found that when temperatures were maintained above $33°C$ the dominant species of the algal community were species of blue-greens; some green algae were quite common and the diatom flora was almost non-existent except for a very few species, most of which had small populations.

Changes in temperature rhythms have effects upon the reproduction of organisms. The molts of many aquatic insects seem to be triggered by changes in temperature. Likewise, reproduction of fish and oysters has been shown to be stimulated by shifts in temperature. In the case of bass in Maryland, it has been reported that shifts in temperature from the lower 60's to the high 60's in the spring of the year bring about reproduction. Likewise, oysters in Chesapeake Bay have been reported to spawn more successfully when the temperature passes from the high 60's to the low 70's in the spring of the year. Numerous studies that have been carried out indicate that freshwater organisms in the temperate zone seem to be adversely affected when the temperature approaches 32°C and may be fatal at 35°C. The duration of exposure affects greatly the reaction of a species to a given temperature.

Thus, with temperature as with other types of pollution, we first see a shift in sizes of population, then kinds of species, and then the elimination of species and the reduction of biomass, with varying degrees of increase in temperature. It is also evident that sudden large changes in temperatures from hot to cold or cold to hot may deleteriously affect various species. The duration of exposure is a very important factor in determining the severity of the effect.

The effects of suspended and settleable solid loads are evidenced by a reduction in numbers of species and sizes of populations and elimination of species such as caddisflies and blackfly larvae, which are filter feeders. The deposition of silt ultimately reduces the diversity of river bed habitats as it destroys the roughness of the bed. If it forms a shifting substrate, it is an unsuitable habitat for most species; thus species diversity and productivity are reduced. In many cases, the suspended solids are rich in organic materials, and as a result, their deposition brings about an increase in bacterial action and reduction in dissolved oxygen in the surface sediments. The fauna shifts from one of burrowing mayflies to one dominated by tubificids and certain species of chironomids.

Suspended solids bring about a reduction in light penetration and thus greatly reduce the primary producers except for those species that are plankton organisms or live on floating debris. As a result, a heavy organic load may not be evidenced by excessive growths of primary producers. This reduction in primary producers may result in a shift in species from those which are herbivores, such as many species of mayflies and caddisflies, and fish, such as gizzard shad, to those that are primarily detritus feeders, such as catfish and carp.

It becomes evident from many studies that the structure of aquatic communities is determined by the numbers of species in the potential species pool and their frequency of invasion; the chemical and physical characteristics of the environment; and the extinction rate of species. Furthermore, it is evident that perturbation of any type, whether natural or man-made, may cause various shifts in kinds of species, relative sizes of populations, and numbers of species. This in turn brings about shifts in predator—prey relationships and interactions of many types between species.

If one wishes to utilize stream communities to determine the effect of pollution, one must carefully select areas for comparison that are ecologically similar — that is, they must include habitats affected by similar rates of current; the areas must include similar bed materials; and the orientation of the areas to the sun as well as the amount of shading must be considered so that the light exposures are not too different. One must collect thoroughly each habitat to obtain all species living in an area. Sizes of populations must be estimated.

We have found that if perturbation does not occur, natural areas support a large number of species belonging to many phyla or groups of organisms and that the number of species and the relative sizes of the populations remain similar over time. Furthermore, we have found that one can much more confidently define the effects of pollution if one bases his conclusions on the behavior of many groups of organisms.

To try to describe these communities by any one single index, such as a diversity index, is rather futile. However, such indices are of value; for example, the Shannon-Weiner diversity index helps determine the evenness of the distribution of individuals in the various species, but it tells little about the biomass that is present and nothing about the kinds of species; and it may or may not reflect the numbers of species present. Therefore it is important, as Levins[8] has pointed out, that one does not rely on any one single index to determine the nature of a community. Various models of communities have been made such as the broken stick model of MacArthur and MacArthur,[9] the truncated normal curve model of Preston,[14] Patrick et al.,[12] and Whittaker;[16] and the dendritic models of Cairns and Kaesler.[3] These models, although more descriptive of the community than an index, do not show the kinds of effects indicated by shifts in types of species. Therefore, one should use many different parameters to correctly describe a community and, by the changes in these various parameters, to determine the effect of pollution.

Biological organisms are very useful in monitoring pollution, because one can determine effects over time and discover the presence of former pollution by analyzing the structure of populations of organisms. Furthermore, organisms integrate all effects that may be deleterious to the growth and reproduction of the species involved. In contrast, chemical analyses more specifically identify pollutants but give one information only for substances analyzed at the point in time when the collections are made. Both approaches are necessary to define pollution effects.

REFERENCES

1. Beeton, A. M. Changes in the environment and biota of the Great Lakes. pp. 150–188 in Eutrophication Symposium. National Academy of Sciences, Washington, D. C., 1969.

2. Brown, E. H., Jr. Survey of the bottom fauna at the mouths of ten Lake Erie, south shore rivers, pp. 156–170 In Lake Erie Pollution Survey. Ohio Dept. Nat. Res. Div. Water, Final Report, 1953.

3. Cairns, J., Jr. and R. L. Kaesler. Cluster analysis of Potomac River survey stations based on protozoan presence-absence data, *Hydrobiologia*, 34(3–4): 414–432 (1969).

4. Cairns, J., Jr. and A. Scheier. The effect upon the pumpkinseed sunfish *Lepomis qibbosus* (Linn.) of chronic exposure to lethal and sublethal concentrations of dieldrin, *Notulae Naturae*, Acad. Nat. Sci. Philadelphia, No. 370, 1964.

5. Carr, J. F., and J. K. Hiltunen. Changes in the bottom fauna of western Lake Erie from 1930–1961, *Limnol. Oceanogr.*, 10: 551–569 (1965).

6. Davis, C. C. Evidence for the eutrophication of Lake Erie from phytoplankton records, *Limnol. Oceanogr.*, 9: 275–283 (1964).

7. Ellis, M. M. Detection and measurement of stream pollution, *U. S. Bur. Fish. Bull.*, 48(22): 365–437 (1937).

8. Levins, R. The strategy of model building in population biology, *American Sci.*, 54: 421–431 (1966).

9. MacArthur, R. H., and J. W. MacArthur. On bird species diversity ... , *Ecology*, 42: 594–598 (1961).

10. Patrick, R. The effects of increasing light and temperature on the structure of diatom communities, *Limnol. Oceanogr.*, 16(2): 405–421 (1971).

11. Patrick, R., B. Crum, and J. Coles. Temperature and manganese as determining factors in the presence of diatom or blue-green algal floras in streams, *Proc. National Acad. Sci.*, 64(2): 472–478 (1969).

12. Patrick, R., M. H. Hohn, and J. H. Wallace. A new method for determinimg the pattern of the diatom flora, *Notulae Naturae*, Acad. Nat. Sci. Philadelphia, No. 259, 1954.

13. Patrick, R., N. A. Roberts, and B. Davis. The effect of changes in pH on the structure of diatom communities, *Notulae Naturae*, Acad. Nat. Sci. Philadelphia, No. 416, 1968.

14. Preston, F. W. The commonness, and rarity, of species, *Ecology*, 29: 254–283 (1948).

15. Tarzwell, C. M., and A. R. Gaufin. Some important biological effects of pollution often disregarded in stram surveys, pp. 295–316 in Proc. 8th Industrial Waste Conf., Purdue Univ. Engineering Bull., 1953.

16. Whittaker, R. H. Dominance and diversity in land plant communities, *Science*, 147: 250–260 (1965).

17. Wilbur, C. G. *The Biological Aspects of Water Pollution*. Charles C Thomas Publ., Springfield, Ill., 1969.

PLANTS AS INDICATORS OF AIR QUALITY

C. Stafford Brandt

Bureau of Air Quality Control
Maryland Environmental Health Administration
610 N. Howard Street
Baltimore, Maryland 21201

Let me start with two opposing syllogisms.

The responses of plants to air pollutants are characteristic of the toxicant. Plants are very sensitive to air pollution. Therefore, plants are excellent indicators of the levels of air pollution.

The responses of plants to air pollutants are dependent upon the specific environmental conditions existing prior to, during, and following exposure to the toxicant. These responses are similar to, or identical with, responses induced by many diseases, insects, and environmental stresses. Therefore, plants are not satisfactory indicators of air pollution.

All these premises are correct. The responses of plants to air pollutants are generally characteristic, but they are not necessarily specific. Similar, if not identical, symptoms are induced by disease, insects, or environmental stresses. Certain plants are generally very sensitive to certain air pollutants, but sensitivity varies greatly among species, varieties, and selections within varieties; and this sensitivity is markedly influenced by environmental conditions prior to, during, and following exposure.

Laying aside the pitfalls of syllogistic reasoning, we can find supporters of either conclusion noted above. Further, we can describe real situations where one or the other conclusion is justified. The problem of using vegetation as an index of air pollution rests upon our ability to describe where any given factual situation lies along the scale of usefulness between these two limiting conclusions.

VISIBLE EFFECTS

Most general reviews [2,3,7,41] of the effects of air pollution on vegetation provide descriptions of the specific visible effects of the pollutants on vegetation. There are now at least three compilations of color plates showing the common effects. A small collection by Hindawi[22] provides a good introduction to the identification of visible injury symptoms. A more extensive collection, edited by Jacobson and Hill[23] provides good pictures and text for the common pollutants. The atlas of Van Haut and Stratmann[43] is limited to the effects of sulfur dioxide. Hindawi[22] and Jacobson and Hill[23] point out some of the confusing factors induced by disease, insects, and stress. Brandt and Heck[3] and, to a lesser extent, Garber[7] and Berge[2] discuss some of the problems of using plants in surveys of air pollution injury.

SULFUR DIOXIDE POLLUTION

While recognition that specific injury symptoms were useful as indicators of air pollution dates back to before the turn of the century,[3,7,41] the approach used by Ruston[36] in 1921 is of interest. He spoke specifically to the use of the plant as an index of smoke pollution. Now, we would probably be inclined to read "sulfur dioxide" for "smoke." Many of his observations do, in fact, relate to smoke and the deposit of soot on leaves. Other of his observations — such as loss of leaves, early senescence, and "characteristic brown blotches" — would now be considered evidence of chronic and acute sulfur dioxide injury. We might even consider some as evidence of photochemical smog. Regardless of the specific causative agent, Ruston did successfully use vegetation as an index of air pollution impact, first around Leeds and then in other industrialized areas.

The general problem of vegetation as an index specifically for sulfur dioxide pollution was faced by McKee and Bieberdorf.[31] They used the classic symptoms of sulfur dioxide to delineate the pollution around Houston. They also used sulphate analysis of leaf tissue to confirm and to give quantitative values to their observations. Confirmation of the observed symptoms of suspected sulfur dioxide injury is a serious problem. For McKee and Bieberdorf, sulfate analysis provided fairly good confirmation.

The question of identification of observed visible symptoms on vegetation, and of the confirmation of the causative agent, has been thoroughly discussed.[13,20,23,38] The answers are mixed as far as sulfur dioxide pollution and sulfate accumulation are concerned. For the acute fumigation, sulfate analysis is rather useless for confirmation.[11,41] When the problem is chronic injury occurring as a result of low concentration of sulfur dioxide for long periods of time, the amount of sulfate in leaf tissue may provide confirmatory evidence.[11,28]

FLUORIDE POLLUTION

The fluoride problem is unique, and the use of vegetation as an index of fluoride pollution,[1] as well as a monitor,[30,32] is well established. Because the limit of fluoride in the diet of cattle is critical,[32] the amount of fluoride in plant tissue is often of more importance than visible injury symptoms in delineating the extent of fluoride pollution.

PHOTOCHEMICAL SMOG POLLUTION

In the early days of the smog problem in the Los Angeles basin, vegetation was almost the only index of smog available.[24,33] Vegetation has proved to be an equally effective demonstrator of photochemical smogs 2500 miles east of Los Angeles.[21]

The problems of using vegetation as indicators and measures of photochemical smog are similar to the problems with any of the other phytotoxicants.[14,16] The visible symptoms are characteristic but by no means specific. Sensitivity varies markedly among selections and with environmental conditions. Tobacco weather fleck appears to be quite specific for ozone, one of the phytotoxicants in smog,[3] and the occurrence of weather fleck in the field correlates very well with measured ozone (or oxidant) concentrations.[29] In an effort to overcome some of the problems. the group at Cincinnati,[15] with which I was associated for several years, devised a system aimed at reducing some of the variables in culture, exposure, and environment. Measures of the degree of leaf injury correlate very well with chemical measures of oxidant under a variety of conditions.

GROWTH EFFECTS

Many of the responses of vegetation to air pollution are subtle. Early senescence is recognized as a common response to chronic sulfur dioxide[3,41] and of photochemical smogs.[3,14] Heggestad[17,18] has reported on reduced growth due to smog or general air pollution in studies conducted in filtered versus unfiltered air at Beltsville, Maryland. The response has been noted on many more species than he has reported.[19] Treshow[42] has discussed the impact of air pollution on plant populations, suggesting that air pollution does cause a shift in distribution of species. Knabe[25] has correlated the distribution of Scots pine in the Ruhr region with the distribution of sulfur dioxide concentrations.

A very large number of workers[5,6,8-10,12,26,27,34,35,37,39,40] from many sections of the world have used lichens and some of the epiphytic mosses as indices of air pollution. Several[10,12,39] have devised semi-quantitative scales based on values assigned to the frequency of occurrence of certain species. Schönbeck,[37] using a technique of Brodo,[4] transplants lichens to specific exposure sites in a region and measures their growth or, more properly, their lack of growth and decline. There is good agreement among workers that the lichen indexes are measures of air pollution and not other environmental changes. There is also agreement that lichen distributions are an index of the extent of the dust, dirt, grime, and sulfur dioxide complex of urban and industrial areas. As yet, we do not have clear evidence that the lichen also serves as an index of photochemical smog.

USES AND NEEDS

Plants are indicators of air pollution. As with any indicator, their usefulness depends upon how we interpret the indications received.

Field surveys of vegetation are a useful, and often necessary, tool in air pollution studies. Often such surveys are in the nature of confirming the obvious. We know that an area has air pollution, and we attempt to use the symptoms on plants in the field to delineate the extent of the pollution. Sometimes the observations of injury are useful for stimulating public action. Discolored, shriveled leaves and bare branches can be, politically, rather photogenic.

The use of field vegetation has limitations. First, to read correctly the visible symptoms and to sort them out from all the other factors that could produce the symptoms requires a competent observer. Also, the observer must be in the field to read the symptoms at the proper time. Second, sensitive vegetation must occur with a reasonable distribution over the area. If these conditions are met, the survey will confirm the obvious. Further, it should delineate the area where vegetation is affected and provide some estimate of the severity of the effects. Since the actual vegetation normally growing in the area is affected, the survey may provide some estimates of the impact of air pollution upon agricultural production, property values, and land use.

The use of some groups of plants, such as lichens or mosses, has similar limitations. The response, presence or absence of a species, is, perhaps, more easily read than visible symptoms of injury. The requirement for a competent observer, in this case one who can identify species even if modified by air pollution, is still the same. Because of the ubiquity of lichens, the distribution problem is less. As a result, this technique gives an excellent index of the extent and intensity of air pollution of the smoke—sulfur dioxide type over an area. When we try to extrapolate the defined impact upon lichens to possible impact upon productivity, value, or land use, we run into difficulties.

If we transplant lichens, or if we use a bioassay technique such as a specific species of tobacco exposed under standardized conditions, we still have difficulties. We improve specificity. We may improve our ability to quantitate response. We reduce the degree of competency required in our field personnel. All of these points are to the good. We do not improve our ability to extrapolate from a specific response to a general response index that measures productivity, value, or land use restrictions.

To borrow from the jargon of the computer field, the first and second generations of plants as indices of air pollution are getting old. We need a third generation.

We are no longer killing vegetation with sulfur dioxide in the manner of Ducktown, Tennessee. Instead, the major problem is chronic injury expressed as generalized chlorosis, early senescence, and poor growth. The acute stage of Los Angeles smog may still be there, but the important problem is the widespread generalized chlorosis, early senescence, and reduced growth. For these problems, field surveys are almost useless. How do you identify poor growth, early old age, or even a generalized chlorosis? Even the annual bluegrass and pinto bean bioassays that helped identify smog give no suggestion of the extent of growth reduction noted by Noble.[33] The tobacco bioassay provides no better measure of the growth reduction reported by Heggestad.[19]

I can suggest one possible form of third generation index generator; there are undoubtedly others. My generator consists of lichens exposed in some manner such as that suggested by Schönbeck[37] and tobacco exposed under controlled conditions such as suggested by Heck et al.[15] These constitute the sensors, one for the dust–dirt–grime–SO_2 complex and one for the smog complex. So far there is nothing new or unique. However, I believe we are close to being able to derive a set of factors that, when applied to the readout of our sensors, will yield indices relatable to productivity. If Heggestad or Heck or someone else can relate observed growth differences under filtered vs unfiltered air to the readout of a biological smog sensor, we have part of our complete index generator. If Schönbeck or Guderian or someone else could do the same type of study in the Ruhr, we have the other part of the index generator.

With some imagination, some hard work, and, of course, some money, I believe we can develop a system of biologically generated indices of air quality that will provide scalar estimates of general vegetative productivity.

REFERENCES

1. Adams, D. F., et al. Relationship of Atmospheric Fluoride Levels and Injury Indexes on Gladiolus and Ponderosa Pine, *Agr. Food Chem.*, 4: 64–66 (1956).

2. Berge, H. Immissionsschäden: Gas–, Rauch–, und Staubschäden, in *Sorauer, Handbuch der Pflanzenkrankheiten*, Bd. 1, Paul Parey, Berlin and Hamburg, 1970.

3. Brandt, C. S., and W. W. Heck. Effects of air pollutants on vegetation, pp. 401–443 in *Air Pollution*, 2nd Ed., Vol. I, A. C. Stern (ed.). Academic Press, New York, 1968.

4. Brodo, I. M. Transplant experiments with corticolous lichens using a new technique, *Ecology*, 42: 838–841 (1961).

5. DeSloover, J., and F. LeBlanc. Mapping of atmospheric pollution on the basis of lichen sensitivity, *Proc. Symp. Recent Adv. Trop. Ecol.*, 1968: 42–56.

6. Domros, M. Flechten als Indikator von Luftverunreinigung und Stadtklima, *Städtehygiene*, 18 (2): 33–39 (1967).

7. Garber, K. *Luftverunreinigungen und ihre Wirkungen*. Gebruder Borntrager, Berlin, 1967.

8. Gilbert, O. L. Lichens as indicators of air pollution in the Tyne valley, pp. 35–47 in *Ecology and the Industrial Society*, Oxford, 1965.

9. Gilbert, O. L. Bryophytes as indicators of air pollution in the Tyne valley, *New Phytol.*, 67: 15–30 (1968).

10. Gilbert, O. L. A biological scale for the estimation of sulphur dioxide pollution, *New Phytol.*, 69: 629–634 (1970).

11. Guderian, R. Untersuchunger über quantitative Beziehungen zwischen dem Schwefelgehalt von Pflanzen dem Schwefeldioxidgehalt der Luft (im drei

Teile), *Z. Pflanzenkrankh. (Pflanzenpathol.) Pflanzenschutz*, 77: 200–220; 289–308; 387–399 (1970).

12. Hawksworth, D. L., and F. Rose. Qualitative scale for estimating sulphur dioxide air pollution in England and Wales, using epiphytic lichens, *Nature*, 227: 145–148 (1970).

13. Heck, W. W. The use of plants as indicators of air pollution, *Int. J. Air. Water Poll.* 10: 99–111 (1966).

14. Heck, W. W. The use of plants as sensitive indicators of photochemical air pollution, Proceedings of the International Symposium on Identification and Measurement of Environmental Pollutants, June, 1971, Montreal, Ontario, 1972.

15. Heck, W. W., F. L. Fox, C. S. Brandt, and J. A. Dunning. *Tobacco, a sensitive monitor for photochemical air pollution*. Nat. Air Poll. Cont. Adm. Pub. No. AP-55 (1969).

16. Heck, W. W., and A. S. Heagle. Measurement of photochemical air pollution with a sensitive monitoring plant, *J. Air Pol. Cont. Assoc.*, 20: 97–99 (1970).

17. Heggestad, H. E. Discussion of paper by Dr. Taylor (Effects of oxidant air pollutants), *J. Occup. Med.*, 10: 492–96 (1968).

18. Heggestad, H. E. Variations in response of potato cultivars to air pollution, *Phytopathology*, 60: 1015 (1970).

19. Heggestad, H. E. Personal communication, 1971.

20. Heggestad, H. E., and W. W. Heck. Nature, extent and variation of plant response to air pollutants, *Adv. Agronomy*, 23: 111–145 (1971).

21. Hindawi, I. J. The product of photochemical reactions as reflected on vegetation grown in Brooklyn Botanical gardens, Willowbrook, N. Y., Roselle and Bayonne, New Jersey. Presented at Cooper Union Colloquium, New York City, April, 1967.

22. Hindawi, I. J. *Air Pollution Injury to Vegetation*. Nat. Air Poll. Cont. Adm., Pub. No. AP-71 (1970).

23. Jacobson, J. S., and A. C. Hill (eds.). *Recognition of air pollution injury to vegetation: a pictorial atlas*. Air Pollution Control Assn., Pittsburgh, Pa. (1970).

24. Juhren, M., W. M. Noble, and F. W. Went. The standardization of *Poa Annua* as an indicator of smog concentrations. I. Effects of temperature, photoperiod and light intensity during growth of the test plants, *Plant Phys.*, 32: 576–586 (1957).

25. Knabe, W. Kiefernwaldnerbreitung und Schwefeldioxid– Immissionen im Ruhrgebiet, *Staub,* 30: 32–35 (1970).

26. LeBlanc, F. Possibilities and methods for air pollution on the basis of lichen sensitivity, *Mitterlungen der Forstlichen Bundes–Versuchsanstalt, Wien,* 92: 103–126 (1971).

27. LeBlanc, F. and J. DeSloover. Relation between industrialization and the distribution and growth of epiphytic lichen and mosses in Montreal, *Can. J. Bot.*, 48: 1485–1496 (1970).

28. Lihnell, D. Sulfate contents of tree leaves as an indicator of SO_2 air pollution in industrial areas, pp. 341–353 in *Proc. First Europ. Congr. Infl. Air Poll. on Plants and Animals*, Wageningen (PUDOC), 1969.
29. Macdowell, F. D. H., E. J. Mukammal, and A. I. W. Cole. Direct correlation of air polluting ozone and tobacco weather-fleck, *Can. J. Plant Sci.*, 44: 410–417 (1964).
30. MacIntire, W. H., M. A. Hardison, and D. R. McKenzie. Spanish moss and filter paper exposures for detection of air-borne fluorides, *Agr. Food Chem.*, 4: 613–620 (1956).
31. McKee, H. C., and F. W. Bieberdorf. Vegetation symptoms as a measure of air pollution, *J. Air Pol. Cont. Assoc.*, 10: 222–225 (1960).
32. National Research Council, Committee on Biologic Effects of Atmospheric Pollutants, *Fluorides*, National Academy of Sciences, Washington, D. C. 1971.
33. Noble, W. M., and L. A. Wright. Air pollution with relation to agronomic crops: II. A bio-assay approach to the study of air pollution, *Agron. J.*, 50: 551–553 (1958).
34. Pyatt, F. B. Lichens as indicators of air pollution in a steel producing town in South Wales, *Environ. Pollut.*, 1: 45–56 (1970).
35. Rose, F. Lichens as pollution indicators, *Your Environment (London)*, 1: 185–189 (1970).
36. Ruston, A. G. The plant as an index of smoke pollution, *Ann. Appl. Biol.*, 7: 390–402 (1921).
37. Schönbeck, H. Eine Methode zur Erfassung der biologischen Wirkung von Luftverunreinigungen durch transplantierte Flechten, *Staub*, 29 (1): 14–11 (1969).
38. Schönbeck, H., M. Buch, H. van Haut, and G. Scholl. Biologische Messverfahren für Luftverunreinigungen, *VDI Berichte*, 149: 225–236 (1970).
39. Skye, E. Användandet av lavar som indikatorer och testorganismer pa luftföroreningar, *Nordish Hyg. Tidsk.*, 69: 115–134 (1969).
40. Skye, E., and I. Hallberg. Changes in the lichen flora following air pollution, *Oikos*, 20: 547–552 (1969).
41. Thomas, M. D. Effects of air pollution on plants, pp. 233–278 in *Air Pollution*, WHO monograph Ser. 46. World Health Organization, Geneva, 1961.
42. Treshow, M. The impact of air pollutants on plant populations, *Phytopath.*, 58: 1108–1113 (1968).
43. Van Haut, H., and H. Stratmann. *Farbtafelatlas über Schwefeldioxid–Wirkungen an Pflanzen.* W. Gerardet, Essen, 1970.

BIOCHEMICAL INDICATORS
OF ENVIRONMENTAL POLLUTION*

Gerald Goldstein

Leader, Analytical Biochemistry Group
Analytical Chemistry Division
Oak Ridge National Laboratory[†]
Oak Ridge, Tennessee 37830

The practical use of biological indicators to monitor environmental quality has a long history dating back to the miner's canary; to the recognition[4] about 100 years ago of the effect of sulfur dioxide on the vegetation surrounding smelters; and to the observation,[19] around the turn of the century, of the effect of pollutants on the population of flora and fauna living in natural waters. Since that time, our knowledge of the biological effects of the various environmental pollutants has increased enormously, and monitoring schemes employing biological indicator organisms have been proposed and are in fact in daily use.[1] There are certain intrinsic advantages in biological indicators as compared to chemical analysis for individual compounds. Biological indicators are screening agents in that they respond to many different compounds, and they are integrating devices in that they show the cumulative effects over a period of time or over some spatial area; but their primary advantage is that the bio-indicator directly measures the property that we are really interested in — is there something in the air or water that is harmful to life?

The number and kind of biological indicators that may be useful is very large indeed, and it is convenient to subdivide them into a hierarchy as shown in Table 1. One can create four classifications which are, in order of decreasing biological complexity: organisms, tissues and organs, cell cultures, and cell-free preparations. The meaning of the first three classes is obvious. Cell-free preparations will include everything from the most complex mixture containing all the soluble cell components to the simplest, a single enzyme in solution. Since the first three classes are organized living systems, any effects that one observes can reasonably be considered *in vivo* effects, while tests with cell-free preparations will show only *in vitro* effects. An indicator that is an organism is quite clearly a biological indicator. Observations of a cell-free system can only give biochemical information; cell-free preparations can be termed biochemical indicators. I want to emphasize the distinction between biological and biochemical indicators. One

---

*Research sponsored by the National Science Foundation RANN Program
[†]Operated by Union Carbide Corporation for the U.S. Atomic Energy Commission.

Table 1. Hierarchy of biological indicators

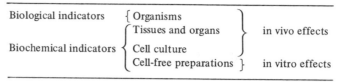

can use the two intermediate classifications — tissues and organs, and cell cultures — in both ways, observing biological effects such as growth and death or biochemical effects such as enzymic or chemical changes.

It should be recognized that the further one gets down the hierarchy away from the organism level, the less reliable any test will be as a measure of harmfulness to life. Substances that inhibit some biochemical process *in vitro* may not be significantly toxic because they do not enter into living systems under ordinary circumstances (because of cellular or tissue barriers) or because they may be very efficiently detoxified internally by enzymatic action. On the other hand, *in vitro* assays in particular are easy to do and are inexpensive compared to experiments using living organisms, and it may be feasible for communities that lack the financial and technical resources to effectively use a biological indicator scheme to use a biochemical indicator where the pertinent tests might be done in the clinical laboratory of the local hospital for example. Finally it is quite possible that observations of biochemical indicators can in some cases be related to what is probably the most pressing and least well understood of our environmental problems — what is the long term effect of chronic exposure to low concentrations of hazardous substances? A conventional experiment to answer this kind of question might easily involve billions of test animals, scientific man-hours, and dollars; and even if it were possible to carry out such an experiment, there is still no guarantee that the data obtained from test animals can be applied to human beings. Any additional information that will shed some light on this problem is sorely needed, and it is not unreasonable to expect that observations of biochemical effects on appropriate biochemical indicators can at least generate some insight into the nature of the biochemical lesions that might occur on long term exposure and whether or not one might expect to find a threshold concentration level.

Consequently I propose to review, in a very brief and selective way, some of the studies that have been made that are related to the concept of biochemical indicators as I have defined them and, in particular, to those which would appear to be likely candidates for monitoring schemes.

Several kinds of tissues and organs are potentially useful as indicators including plant tissues, various animal tissues, and blood.

The little work that has been done with plant tissues is almost entirely due to Ordin and his associates,[28-32] who observed the effect of exposure to peroxyacetyl nitrate (PAN), ozone, and fluoride on the biosynthesis of cellulose

Table 2. Effect of pesticides on fish-brain cholinesterase

Fish	Pesticide	Concentration (mg/1)	Exposure time	AChE Activity (% of normal)
Minnow	Parathion	0.1	7h	85
Goldfish	Parathion	0.1	7h	75
Bluegill	Parathion	0.1	7h	15
Bluegill	Parathion	0.001	15d	30
Bluegill	Malathion	0.001	15d	80
Bluegill	Guthion	0.001	15d	10

in isolated oat cell wall sections and observed inhibition in all cases. The inhibition was attributed to the inactivation of two enzymes — phosphoglucomutase and cellulose synthetase — both of which were inactivated *in vitro* by PAN. Although these particular experiments are probably too cumbersome for routine application, they do illustrate the ability of isolated plant tissues to act as monitors, responding to several different pollutants and integrating the effects over a period of time.

Considerably more work has been done with animal tissues, the bulk of it involving the detection of pesticides in natural waters by measurement of acetylcholinesterase enzyme activity in the brains of indigenous fish. The inhibition of cholinesterase enzymes by organophosphate and carbamate compounds is of course well documented, and exposure of fish to these compounds reduces the brain cholinesterase activity, with death occurring when the activity level falls to 30 to 60% of that of nonexposed fish. Table 2 shows some data taken from reports by Weiss and his coworkers[46-49] on the effect of some organophosphorous pesticides on the brain cholinesterase activity of different fish. The first three sets of data show the inhibition of fathead minnow, goldfish, and bluegill cholinesterase by exposure to 0.1 mg/liter of parathion for 7 hours; the residual cholinesterase activity of the surviving fish was 85, 75, and 15%, respectively. This is obviously a toxic dose to sensitive species such as the bluegill. The last three sets of data show the effect of a 15-day exposure to 1 ppb of parathion, malathion, and guthion on bluegill cholinesterase, the residual activities being 30, 80, and 10% respectively. This is chronic, sub-lethal exposure, and from such data Weiss has concluded that as little as 0.1 ppb of pesticide could be detected by exposure of a suitably sensitive species of fish for 30 days. The idea of using fish-brain cholinesterase as a monitor for pesticides has in fact been successfully applied,[3,17,50] and it has even been suggested that a maximum 10% depression of brain cholinesterase activity in indigenous fish be adopted as one criterion of water quality.[27]

Other animal tissues have been used, or could be used, as indicators in a variety of ways. A few examples are shown in Table 3. In the first case,[18] fish were exposed to six toxic metals — Pb, Cu, Hg, Be, Cd, and Ag — in

Table 3. Animal tissues as biochemical indicators

Tissue	Pollutant, amount, and exposure time	Observed effects
Fish liver	Six toxic metals, 0.04 ppm (Ag)-188 ppm (Pb), 96 hrs	Reduced levels of alkaline phosphatase, acid phosphatase, xanthine oxidase, and catalase
Rat liver	DDT, 5–10 ppm, 4 weeks,	Reduced level of glucose-6-phosphate dehydrogenase
Guinea pig lung, liver, spleen, kidney	NO_2, 15 ppm, 10 weeks	Elevated lactic dehydrogenase and aldolase in most tissues

concentrations ranging from 0.04 ppm for Ag to 188 ppm for Pb, for a period of 96 hours. The activities of several of their liver enzymes were then measured and compared with controls. The avowed purpose of this study was to determine whether liver enzyme activities could be used as indicators of sublethal exposure of fish to toxic metals. Reduced levels of alkaline phosphatase, acid phosphatase, xanthine oxidase, and catalase were observed. Catalase activity, in particular, was reduced by all the metals tested.

In the second example,[43] rats were fed diets containing various amounts of DDT for a period of four weeks, following which the activity of glucose-6-phosphate dehydrogenase was measured in the liver. The interesting feature of this work is that a dietary level of only 5 ppm of DDT was sufficient to induce a significant reduction of enzyme activity. Although 5 ppm is probably too high to be considered a chronic exposure, it does point out what one might look for in evaluating chronic effects of DDT and other chlorinated hydrocarbons.

In the last example,[2] guinea pigs were exposed to 15 ppm of NO_2 continuously for 10 weeks; and lung, liver, spleen, and kidney tissues were subsequently examined. Lactic dehydrogenase and aldolase enzyme levels were elevated in most tissues. This is a good illustration that chronic inhalation of air pollutants can cause alterations in tissue enzyme activity levels which can be detected and perhaps used as sensitive indicators.

There are of course many more examples of studies of this kind in the physiological and toxicological literature which could, with a little ingenuity, be applied as monitoring schemes.

Using blood as a test tissue to monitor exposure to pollutants is an attractive idea, particularly if one would like to screen large populations. It is of course very easy to obtain blood samples without the need of sacrificing animals, thereby offering us the possibility of using ourselves as indicator species. This may sound a little grim, but it should be recognized that we are, after all, the ultimate indicators anyway, in the sense that the first responsibility of any monitoring or control program is the protection of human health. A few examples of how one might use blood chemistries in pollution monitoring are shown in Table 4.

The first example involves the measurement of cholinesterase enzyme activity in human blood plasma and erythrocytes as an indicator of exposure to organophosphorous and carbamate pesticides,[1,2] a test that has been used for years in the control of occupational exposure. The difficulty of using this test, or any other blood chemistry test, to survey general populations is that there is a substantial individual variation in enzyme activity in unexposed individuals; and even in the same individual, the activity will change with time. In the occupational context, the normal range for an individual can be established by repeated measurements over a long period of time, and a departure from this range can be detected. Long-term testing is probably not feasible for large populations, but based on what data is available, it has been suggested that an individual with red cell cholinesterase activity 20% below the population average, or a plasma activity 33% below, has probably been exposed to a cholinesterase inhibitor. The second example in Table 4 is the effect of exposure to 1 ppb of endrin, a chlorinated hydrocarbon pesticide, on the blood chemistry of a marine fish, the northern puffer.[6] The blood concentrations of sodium, potassium, calcium, and cholesterol as well were consistently higher in experimental fish than controls, probably reflecting endrin-induced liver damage. Although very little is known about the normal blood chemistry values of fish, it is reasonably easy to establish control populations for comparison purposes, and in this way to possibly use fish-blood tests to monitor the marine environment. In the third example in this table, human blood was again tested, in this case for a decrease in the enzyme δ-aminolevulinic acid dehydrase due to exposure to lead.[16] Absorption of lead in the human body gives rise to a variety of biological and biochemical manifestations, but the biochemical effects are of course evident long before clinical symptoms are present. The activity of the enzyme δ-aminolevulinic acid dehydrase, which is involved in the biosynthesis of hemoglobin, is in particular extremely sensitive to lead and is demonstrably inhibited by blood-lead concentrations as low as 5 to 25 μg per 100 ml of blood.

Table 4. Blood as a biochemical indicator

Blood source	Pollutant	Observed Effect
Human blood	Organophosphorous and Carbamate pesticides	Decreased blood cholinesterase activity
Marine fish	Endrin (1 ppb, 96 hours)	Elevated serum Na, K, and Ca
Human blood	Lead (5–50 μg per 100 ml)	Decreased activity of δ-aminolevulinic acid dehydrase

This, as it happens, is the range of blood-lead levels found in the general population which has never been occupationally exposed. Besides providing a good biochemical indicator of exposure to lead, the demonstration that present levels of environmental contamination with lead produce a measurable bio-chemical effect in man bear upon the point that was made earlier — that in evaluating the potential effects of long term exposures, biochemical information is bound to be very useful.

Turning to the next lower level of the biological hierarchy, cell cultures, a few of the things that have been done are shown in Table 5. Although cultured cells have been shown to be responsive to a number of environmental pollutants, they have not really been exploited as biochemical indicators since it is usually biological effects that have been looked at. Briefly, nitrogen dioxide has been tested against mouse liver[34] and rabbit and rat lung cell cultures.[39] In the first case, increased cell mortality was observed, and in the latter, a reduced oxygen consumption, reflecting reduced cellular metabolism. Sulfur dioxide inhibits the growth of mouse liver cells[42] and damages human lymphocytes in culture[40] as reflected by reduced cell growth, DNA synthesis, and mitotic indices; and in addition chromosome abnormalities were observed. Ozone inhibits the growth of human strain L cells.[33] In several studies, various pesticides were found to have a variety of effects on mammalian cell cultures.[7-10,20,22] In addition to cytotoxicity and growth inhibition, treated cultures were more susceptible to other toxins and also to virus infections. The last item in the table is particularly interesting. Rachlin and Perlmutter[36] compared the response of a whole animal (the fathead minnow) and also its cells in culture to zinc in solution. They found that the biologically safe concentration, as determined by whole animal toxicity tests, actually reduced the mitotic index of the cell cultures by approximately 50%, suggesting that embryos and new-hatched fish, with rapidly dividing cells, will be more sensitive to zinc than adults. Consequently tissue culture tests may provide more reliable criteria of biologically safe concentrations than whole animal tests if only the adult is tested.

Table 5. Cell cultures as biochemical indicators

Pollutant	Cell culture	Observed effect
NO_2	Mouse liver, rabbit and rat lung	Cell mortality and reduced oxygen consumption
SO_2	Mouse liver, human lymphocytes	Growth inhibition and lymphocyte damage
Ozone	Human strain L	Growth inhibition
Pesticides	Human: liver, HeLa: mouse: liver, fibroblast	Growth inhibition
Zinc	Minnow epithelial	Reduced mitotic index

Finally, I want to discuss the potential value of the lowest level of our hierarchy, cell-free biochemical systems, as indicators. Focussing on the simplest form of cell-free system, a single enzyme in solution, it turns out that we are simply looking from a different angle at a field of analytical chemistry called enzymatic methods of analysis and can take advantage of the information that has been developed over the years. Enzymatic methods have two important applications in analytical chemistry. First, enzymatic methods are employed routinely in clinical testing laboratories either for the analysis of enzyme substrates – the estimation of glucose using glucose oxidase or hexokinase is probably the most familiar example – or for analysis of blood and serum enzymes themselves. I do not think that we will find many pollutants to be enzyme substrates; so substrate analysis is probably not very important in pollution monitoring. However, because enormous numbers of assays for enzymes also need to be done in clinical laboratories, a variety of automated instrumentation has been developed for measuring enzyme activities, and that is something worth keeping in mind. The second important application of enzymatic methods is the determination of pesticide residues in agricultural products and foodstuffs by *in vitro* inhibition of cholinesterase enzymes.[11] In fact, in some cases the enzyme inhibition assay is the official test required by the Food and Drug Administration, the organization responsible for setting the legal tolerance limits. It is evident that this assay can easily be adapted to environmental testing, and this has been done as I will mention later.

The primary difficulty with enzymatic inhibition methods from the point of view of the analytical chemist is their relative lack of selectivity. Most enzymes are extremely unstable molecules and can be denatured or inhibited by a large number of compounds, even by sufficiently high concentrations of something as common as sodium chloride. If one is attempting to analyze for a single compound, this is bad news; but in the context of indicators and monitoring, I am inclined to consider it more of an advantage than a disadvantage. An indicator, after all, is meant for screening not as a method of assay for a particular compound, and it is just as important to detect unusually high concentrations of common compounds as it is to detect rare compounds.

Table 6. Effect of metal ions on yeast alcohol dehydrogenase

Metal ion	Concentration for 50% inhibition (M)
Hg(II)	7.5×10^{-9}
Ag(I)	6.5×10^{-9}
Cd(II)	1.0×10^{-7}
Cu(II)	7.5×10^{-6}
Co(II), Bi(III)	1.3×10^{-4}
Be, Pb, Ca, Sr, Ba, Mg, Mn(II)	(1.3×10^{-4}, no inhibition)

Furthermore the sensitivity of an enzyme to different inhibitors can vary enormously, and a good example is the enzyme yeast alcohol dehydrogenase as shown in Table 6. Townshend and Vaughan[45] determined the concentrations of various metal ions required to inhibit the enzyme activity by 50%. The concentrations ranged from about 7×10^{-9} M for silver and mercuric ions to about 10^{-4} for cobaltous and bismuth ions — a difference of five orders of magnitude in sensitivity — while 10^{-4} M concentrations of beryllium, lead, calcium, strontium, barium, magnesium, and manganous ions did not inhibit the enzyme at all. Consequently, I can forsee using yeast alcohol dehydrogenase as an indicator for silver and mercuric ions in water, for instance, without having to worry about the concentrations of the other elements normally present. Of course if there were something unusual about the water — it might be very hard, or saline, or grossly polluted — it is quite possible that the total concentration of the common elements sodium, calcium, and magnesium could be high enough to inhibit the enzyme. But isn't this something that we would want to know? Our indicator would signal an abnormal condition, which is precisely its purpose. Yeast alcohol dehydrogenase is by no means unique. Most enzymes are extraordinarily sensitive to certain classes of compounds and insensitive to others, depending on the particular composition and structure of the enzyme.

Enzymatic methods of analysis have been described for other metals also, and a few examples are shown in Table 7. Alkaline phosphatase is inhibited by micromolar concentrations of beryllium and bismuth.[15] This is one of the few enzymes that are sensitive to beryllium, and beryllium and bismuth are the only cations that are strong inhibitors. Micromolar concentrations of cupric and ferrous ions have been analyzed by inhibition of hyaluronidase,[14] while millimolar concentrations of other metal ions, including lead, mercury, and nickel, have no significant effect. Peroxidase would be less selective as an indicator since it is inhibited by several cations, most strongly by manganous and cobaltous ions,[13] which inhibit at 10^{-4} to 10^{-5} M concentrations. Millimolar concentrations of cupric, ferrous, and lead ions also inhibit. Urease, like yeast alcohol dehydrogenase, is inhibited by most heavy metal ions but most strongly by silver and mercuric ions.[44] Nanogram quantities of these two elements are

Table 7. Enzyme indicators for cations

Enzyme	Cations	Concentration Range (M)
Alkaline phosphatase	Be(II), Bi(III)	$10^{-5} - 10^{-6}$
Hyaluronidase	Fe(II), Cu(II)	$10^{-5} - 10^{-6}$
Peroxidase	Mn(II), Co(II)	$10^{-4} - 10^{-5}$
	Cu(II), Fe(II), Pb(II)	$10^{-3} - 10^{-4}$
Urease	Ag(I), Hg(II),	$10^{-7} - 10^{-8}$
	other heavy metals	$10^{-6} - 10^{-7}$

easily detected, as well as milligram quantities of copper, cadmium, nickel, and so forth. None of these enzymes is affected at all by reasonable concentrations of sodium, potassium, calcium, and magnesium, and all are commercially available.

Enzymes can be inhibited by anionic substances as well as by cations, and several analytical methods have been proposed, three of which are shown in Table 8. Peroxidase is strongly inhibited by sulfide ions, which can be detected at the 10^{-4} to 10^{-5} M concentration level in the presence of chloride, nitrate, and most other anions.[13] Cyanide ion also inhibits peroxidase at 10^{-3} to 10^{-4} M concentrations. Hyaluronidase is a more sensitive indicator for cyanide. As little as 0.1 μg of cyanide per ml of sample can be detected.[14] Fluoride ion is an extremely powerful inhibitor of liver esterase enzymes, and an assay method based upon inhibition has been employed to estimate nanogram quantities of fluoride in complex samples such as blood, urine, and tooth enamel,[21,23,24] since most other anions do not inhibit the enzyme.

I did mention earlier that *in vitro* cholinesterase inhibition has been adapted to environmental monitoring, specifically for the detection of organophosphate pesticides in water. In Table 9 three enzymes are listed – cholinesterase, catalase, and phosphomonoesterase – that have been suggested as indicators of water pollution. In 1967, David and Malaney[5] proposed a cholinesterase inhibition test as a new parameter of water pollution in drinking water and proceeded to prove their point by detecting cholinesterase inhibitors in a municipal water supply. To the best of my knowledge, municipal anthorities have expressed no interest. A field testing kit for organophosphate pesticides in ground and surface waters has been successfully developed using freeze-dried

Table 8. Enzyme indicators for anions

Enzyme	Anions	Concentration Range (M)
Peroxidase	S=	$10^{-4} - 10^{-5}$
	CN⁻	$10^{-3} - 10^{-4}$
Hyaluronidase	CN⁻	$10^{-4} - 10^{-5}$
Liver esterase	F⁻	$10^{-7} - 10^{-8}$

Table 9. Biochemical indicators of water pollution

Indicator enzymes	Pollutant
Cholinesterase	Pesticides
Catalase	Microbial content
Phosphomonoesterases	Phytoplankton and bacteria

acetylcholinesterase.[51] This is of some significance because the organo-phosphorous pesticides are degradable, and a limited number of fixed monitoring stations is not likely to give a true picture of the extent of pesticide contamination. Also, to supplement tests of fish-brain cholinesterase activity that were mentioned earlier, Williams and Sova[50] collected samples of river water and tested them for *in vitro* anticholinesterase activity in an effort to locate the sources of pesticide pollution that the fish-brain tests indicated were present. Of 12 samples collected, only three contained cholinesterase inhibitors, two taken from the effluents from a pesticide plant and the other from a fertilizer plant. Certainly there is abundant evidence for the value of choline-sterase enzyme as an indicator for pesticides in water. In the two other examples in Table 9, the enzyme activities in water were measured directly rather than indirectly as the effect of the water on an enzyme. Enzymes, of course, are associated with life, and a sterile body of water without plant or animal life would contain no enzymes. So one can ask whether enzyme activity in water is indicative of microorganisms. While the same information could be obtained by careful bacteriological examination, no one seriously attempts to do this, and we are usually content with a simple coliform count. Catalase activity has been measured in samples of water extensively contaminated with organic matter from sewage and agricultural runoff and could be correlated fairly closely with microbial content,[41] possibly providing a simple, rapid biochemical test for bacterial contamination and the nutrient status of the water. The activity of free phosphomonoesterase enzymes has been measured in the waters of eutrophic lakes and correlated with the amounts of phytoplankton and bacteria[37,38] and it might therefore serve the same purpose.

Studies have also been made of the effects of air pollutants on cell-free systems, mainly aimed at relating physiological responses to biochemical effects rather than at monitoring for pollutants. Nevertheless a few ideas for indicators do suggest themselves, and possible indicators for peroxyacetyl nitrate, ozone, and sulfur dioxide are shown in Table 10. In each case, a solution containing the enzymes was gassed with the pollutant for various periods of time, and the enzyme activity remaining was measured and compared with a control. Peroxyacetyl nitrate inactivates isocitric dehydrogenase, glucose-6-phosphate

Table 10. Biochemical indicators of air pollution

Pollutant	Indicator enzymes
Peroxyacetyl nitrate	Isocitric dehydrogenase G-6-P dehydrogenase Malic dehydrogenase
Ozone	Papain Glyceraldehyde-3-P dehydrogenase
SO_2	Acetylcholinesterase

dehydrogenase, and malic dehydrogenase within a few minutes.[26] All of these enzymes contain functional sulfhydryl groups, and there is good evidence that peroxyacetyl nitrate oxidizes these groups and that the enzymes then lose their biological activity. Ozone is also a strong oxidant and inactivates papain and glyceraldehyde-3-phosphate dehydrogenase, probably by the same mechanism.[25] Acetylcholinesterase is inhibited by atmospheric sulfur dioxide and also by ozone,[35] although the enzyme is not as sensitive to these compounds as it is to organophosphate pesticides.

Let me conclude simply by emphasizing two points: (1) a case can be made for biochemical indicators of environmental quality to supplement biological indicator organisms and chemical testing methods, and (2) in addition to the monitoring possibilities, the basic biochemical information accrued in the development of biochemical indicators will certainly increase our understanding of the chronic effects of pollutants.

REFERENCES

1. *Standard Methods for the Examination of Water and Wastewater,* 13th Ed., American Public Health Association, Inc., New York, 1971.
2. Buckley, R. D., and O. J. Balchum, Acute and Chronic Exposures to Nitrogen Dioxide, *Arch. Environ. Health* 10: 220 (1965).
3. Coppage, D. L. Characterization of Fish Brain Acetylcholinesterase with an Automated pH Stat for Inhibition Studies, *Bull. Environ. Contam. Toxicol.* 6: 304 (1971).
4. Davenport, S. J., and G. G. Morgis. *Air Pollution, A Bibliography,* Bureau of Mines, Bulletin 537, 1954.
5. Davis, T. J., and G. W. Malaney. Acetylcholinesterase Inhibition – A New Parameter of Water Pollution, *Water Sewage Works* 114: 272 (1967).
6. Eisler, R., and P. H. Edmunds, Effects of Endrin on Blood and Tissue Chemistry of a Marine Fish, *Trans. Amer. Fish. Soc.* 95: 153 (1966).
7. Gabliks, J. Responses of Cell Cultures to Insecticides. II. Chronic Toxicity and Induced Resistance, *Proc. Soc. Exp. Biol. Med.* 120: 168 (1965).
8. Gabliks, J., M. Bantug-Jurilla, and L. Friedman, Responses of Cell Cultures to Insecticides. IV. Relative Toxicity of Several Organophosphates in Mouse Cell Cultures, *Proc. Soc. Exp. Biol. Med.* 125: 1002 (1967).
9. Gabliks, J., and L. Friedman, Responses of Cell Cultures to Insecticides. I. Acute Toxicity to Human Cells, *Proc. Soc. Exp. Biol. Med.* 120: 163 (1965).
10. Gabliks, J., and L. Friedman, Effects of Insecticides on Mammalian Cells and Virus Infections, *Ann. N. Y. Acad. Sci.* 160: 254 (1969).
11. Gage, J. C. Residue Determination by Cholinesterase Inhibition Analysis, in *Advances in Pest Control Research,* Vol. IV, R. L. Metcalf (ed.), Interscience Publishers Inc., New York, 1961, p. 183.

12. Gage, J. C. The Significance of Blood Cholinesterase Activity Measurements, *Residue Rev.* 18: 159 (1967).
13. Guilbault, G. G., P. Brignac, Jr., and M. Zimmer. Homovanillic Acid as a Fluorometric Substrate for Oxidative Enzymes. Analytical Applications of the Peroxidase, Glucose Oxidase, and Xanthine Oxidase Systems, *Anal. Chem.* 40: 190 (1968).
14. Guilbault, G. G., D. N. Kramer, and E. Hackley. Fluorometric Determination of Hyaluronidase and of Cu(II), Fe(II), and Cyanide Ion Inhibitors, *Anal. Biochem.* 18: 241 (1967).
15. Guilbault, G. G., M. H. Sadar, and M. Zimmer. Analytical Applications of The Phosphatase Enzyme System. Determination of Bismuth, Beryllium and Pesticides, *Anal. Chim. Acta* 44: 361 (1969).
16. Hernberg, S., and J. Nikkanen, Enzyme Inhibition by Lead Under Normal Urban Conditons, *Lancet* 1970: 63.
17. Holland, H. T., D. L. Coppage, and P. A. Butler, Use of Fish Brain Acetylcholinesterase to Monitor Pollution by Organophosphorous Pesticides, *Bull Environ. Contam. Toxicol.* 2: 156 (1967).
18. Jackim, E., J. M. Hamlin, and S. Sonis. Effects of Metal Poisoning on Five Liver Enzymes in the Killifish (*Fundulus heteroclitus*), *J. Fish. Res. Bd. Canada* 27: 383 (1970).
19. Kolkwitz, R., and M. Marrson. Ecology of Plant Saprobia, p-47 in *Biology of Water Pollution*, L. E. Keup, W. M. Ingram, and K. M. Mackenthun (eds.), Federal Water Pollution Control Administration, 1967; R. Kolkwitz and M. Marsson, Ecology of Animal Saprobia, p. 85, *ibid*.
20. Li, M. F., and C. Jordan, Use of Spinner Culture Cells to Detect Water Pollution, *J. Fish. Res. Bd. Canada* 26: 1378 (1969).
21. Linde, H. W. Estimation of Small Amounts of Fluoride in Body Fluids, *Anal. Chem.* 31: 2092 (1959).
22. Litterst, C. L., E. P. Lichtenstein, and K. Kajiwara, Effects of Insecticides on Growth of HeLa Cells, *J. Agr. Food Chem.* 17: 1199 (1969).
23. McGaughey, C., and E. C. Stowell. Estimation of a Few Nanograms of Fluoride in Presence of Phosphate by Use of Liver Esterase, *Anal. Chem.* 36: 2344 (1964).
24. McGaughey, C., and E. C. Stowell. The Estimation of Nanogram Levels of Fluoride in Fractions of Milligrams of Tooth Enamel by Means of Liver Esterase, *J. Dental Res.* 45: 76 (1966).
25. Menzel, D. B. Oxidation of Biologically Active Reducing Substances by Ozone, *Arch. Environ. Health* 23: 149 (1971).
26. Mudd, J. B. Enzyme Inactivation by Peroxyacetyl Nitrate, *Arch. Biochem. Biophys.* 102: 59 (1963).
27. Nicholson, H. P. Pesticide Pollution Control, *Science* 158: 871 (1967).
28. Ordin, L. Effect of Peroxyacetyl Nitrate on Growth and Cell Wall Metabolism of Avena Coleoptile Sections, *Plant Physiol.* 37: 603 (1962).

29. Ordin, L., and A. Altman. Inhibition of Phosphoglucomutase Activity in Oat Coleoptiles by Air Pollutants, *Physiol. Plant.* 18: 790 (1965).
30. Ordin, L., M. A. Hall, and M. Katz. Peroxyacetyl Nitrate – Induced Inhibition of Cell Wall Metabolism, *J. Air Pollut. Contr. Ass.* 17: 811 (1967).
31. Ordin, L., and B. P. Skoe. Inhibition of Metabolism in Avena Coleoptile Tissue by Fluoride, *Plant Physiol.* 38: 416 (1963).
32. Ordin, L., and B. P. Skoe. Ozone Effects on Cell Wall Metabolism of Avena Coleoptile Sections, *Plant Physiol.* 39: 751 (1964).
33. Pace, D. M., P. A. Landolt, and B. I. Aftonomos, Effects of Ozone on Cells in Vitro, *Arch. Environ. Health* 18: 165 (1969).
34. Pace, D. M., J. R. Thompson, B. T. Aftonomos, and H.G.O. Holck, The Effects of NO_2 and Salts of NO_2 Upon Established Cell Lines, *Can. J. Biochem. Physiol.* 39: 1247 (1961).
35. P'an, A. Y. S., and Z. Jegier. The Effect of Sulfur Dioxide and Ozone on Acetylcholinesterase, *Arch. Environ. Health* 21: 498 (1970).
36. Rachlin, J. W., and A. Perlmutter. Fish Cells in Culture for Study of Aquatic Toxicants, *Water Res.* 2: 409 (1968).
37. Reichardt, W. Catalytic Mobilization of Phosphate in Lake Water by Cyanophyta, *Hydrobiol.* 38: 377 (1971).
38. Reichardt, W., J. Overbeck, and L. Steubing. Free Dissolved Enzymes in Lake Waters, *Nature (London)* 216: 1345 (1967).
39. Rounds, D. E., and R. F. Bils, Effects of Air Pollutants on Cells in Culture, *Arch. Environ. Health* 10: 251 (1965).
40. Schneider, L. K., and C. A. Calkins, Sulfur Dioxide-Induced Lymphocyte Defects in Human Peripheral Blood Cultures, *Environ. Res.* 3: 473 (1970).
41. Sridhar, M. K. C., and S. C. Pillai. Catalase Activity in Polluted Waters, *Effl. Water Treat. J.* 9: 81 (1969).
42. Thompson, J. R., and D. M. Pace, The Effects of Sulfur Dioxide Upon Established Cell Lines Cultivated In Vitro, *Can J. Biochem. Physiol.* 40: 207 (1962).
43. Tinsley, I. J. DDT Ingestion and Liver Glucose-6-Phosphate Dehydrogenase Activity, *Biochem. Pharmacol.* 14: 847 (1965).
44. Toren, E. C., Jr., and F. J. Burger. Trace Determination of Metal Ion Inhibitors of the Urea-Urease System by a pH-stat Kinetic Method, *Mikrochim. Acta* 1968: 1049.
45. Townshend, A., and A. Vaughan. Applications of Enzyme-Catalysed Reactions in Trace Analysis – VI. Determination of Mercury and Silver by Their Inhibition of Yeast Alcohol Dehydrogenase, *Talanta* 17: 299 (1970).
46. Weiss, C. M. The Determination of Cholinesterase in the Brain Tissue of Three Species of Fresh Water Fish and Its Inactivation in Vivo, *Ecology* 39: 194 (1958).
47. Weiss, C. M. Physiological Effect of Organic Phosphorus Insecticides on Several Species of Fish, *Trans. Amer. Fish. Soc.* 90: 143 (1961).

48. Weiss, C. M. Use of Fish to Detect Organic Insecticides in Water, *J. Water Pollut. Contr. Fed.* 37: 647 (1965).
49. Weiss, C. M., and J. H. Gakstatter. Detection of Pesticides in Water by Biochemical Assay, *J. Water Pollut. Contr. Fed.* 36: 240 (1964).
50. Williams, A. K., and C. R. Sova, Acetylcholinesterase Levels in Brains of Fishes from Polluted Waters, *Bull. Environ. Contam. Toxicol.* 1: 198 (1966).
51. Zweig, G., and J. M. Devine. Determination of Organophosphorous Pesticides in Water, *Residue Rev.* 26: 17 (1969).

BIBLIOGRAPHY

TISSUES AND ORGANS

Animal Tissues

Abou-Donia, M. B., and D. B. Menzel. Fish-Brain Cholinesterase: Its Inhibition by Carbamates and Automatic Assay, *Comp. Biochem. Physiol.* 21: 99 (1967).

Buckley, R. D., and O. J. Balchum. Enzyme Alterations Following Nitrogen Dioxide Exposure, *Arch. Environ. Health* 14: 687 (1967).

Casterline, J. L., Jr., and C. H. Williams. Effect of Pesticide Administration Upon Esterase Activities in Serum and Tissues of Rats Fed Variable Casein Diets, *Toxicol. Appl. Pharmacol.* 14: 266 (1969).

Cross, C. E., A. B. Ibrahim, M. Ahmed, and M. G. Mustafa. Effect of Cadmium Ion on Respiration and ATPase Activity of the Pulmonary Alveolar Macrophage: A Model for the Study of Environmental Interference with Pulmonary Cell Function, *Environ. Res.* 3: 512 (1970).

Gibson, J. R., J. L. Ludke, and D. E. Ferguson. Sources of Error in the Use of Fish-Brain Acetylcholinesterase Activity as a Monitor for Pollution, *Bull. Environ. Contam. Toxicol.* 4: 17 (1969).

Hogan, J. W. Water Temperature as a Source of Variation in Specific Activity of Brain Acetylcholinesterase of Bluegills, *Bull. Environ. Contam. Toxicol.* 5: 347 (1970).

Hogan, J. W., and C. O. Knowles. Some Enzymatic Properties of Brain Acetylcholinesterase from Bluegill and Channel Catfish, *J. Fish. Res. Bd. Canada* 25: 615 (1968).

Ramazzotto, L. J., and L. J. Rappaport. The Effect on Nitrogen Dioxide on Aldolase Enzyme, *Arch. Environ. Health* 22: 379 (1971).

Tinsley, I. J. Ingestion of DDT and Liver Glucose-6-Phosphate Dehydrogenase Activity, *Nature (London)* 202: 1113 (1964).

Weber, C. W., and B. L. Reid. Effect of Dietary Cadmium on Mice, *Toxicol. Appl. Pharmacol.* 14: 420 (1969).

Williams, C. H. β-Glucuronidase Activity in the Serum and Liver of Rats Administered Pesticides and Hepatotoxic Agents, *Toxicol. Appl. Pharmacol.* 14: 283 (1969).

Blood

de Bruin, A. Certain Biological Effects of Lead Upon the Animal Organism, *Arch. Environ. Health* 23: 249 (1971).

Casterline, J. L., Jr., and C. H. Williams. The Detection of Cholinesterase Inhibition in Erythrocytes of Rats Fed Low Levels of the Carbamate Banol, *J. Lab. Clin. Med.* 69: 325 (1967).

Goldstein, B. D., and O. J. Balchum. Effect of Ozone on Lipid Peroxidation in the Red Blood Cell, *Proc. Soc. Exp. Biol. Med.* 126: 356 (1967).

Goldstein, B. D., B. Pearson, C. Lodi, R. D. Buckley, and O. J. Balchum. The Effect of Ozone on Mouse Blood in Vivo, *Arch. Environ. Health* 16: 648 (1968).

Haeger-Aronsen, B., M. Abdulla, and B. I. Fristedt. Effect of Lead on δ-Aminolevulinic Acid Dehydrase Activity in Red Blood Cells, *Arch. Environ. Health* 23: 440 (1971).

Hernberg, S., J. Nikkanen, G. Mellin, and H. Lilius. δ-Aminolevulinic Acid Dehydrase as a Measure of Lead Exposure, *Arch. Environ. Health* 21: 140 (1970).

Hogan, J. W. Some Enzymatic Properties of Plasma Esterases from Channel Catfish (*Ictalurus punctatus*), *J. Fish. Res. Bd. Canada* 28: 613 (1971).

Lane, C. E., and E. D. Scura. Effects of Dieldrin on Glutamic Oxaloacetic Transaminase in *Poecilia latipinna, J. Fish. Res. Bd. Canada* 27: 1869 (1970).

CELL CULTURE

Baker, F. D., and C. F. Tumasonis. Modified Roller Drum Apparatus for Analysing Effects of Pollutant Gases on Tissue Culture Systems, *Atmos. Environ.* 5: 891 (1971).

Johnson, W. J., and S. A. Weiss. Cytotoxicity of Dichlorodiphenylacetic Acid (DDA) Upon Cultured KB and HeLa Cells, and its Reversal by Mevalonic Acid, *Proc. Soc. Exp. Biol. Med.* 124: 1005 (1967).

Rachlin, J. W., and A. Perlmutter. Response of Rainbow Trout Cells in Culture to Selected Concentrations of Zinc Sulfate, *Prog. Fish-Cult.* 31: 94 (1969).

Rounds, D. E. Environmental Influences on Living Cells, *Arch. Environ. Health* 12: 78 (1966).

Rounds, D. E., A. Awa, and C. M. Pomerat. Effect of Automobile Exhaust on Cell Growth in Vitro, *Arch. Environ. Health* 5: 49 (1962).

Sachsenmaier, W., W. Siebs, and T. Tan. Wirkung von Ozon auf Mäuseascitestummerzellen und auf Hühnerfibroblasten in der Gewebekultur, *Z. Krebsforsch.* 67: 113 (1965).

Wilson, B. W., and N. E. Walker. Toxicity of Malathion and Mercaptosuccinate to Growth of Chick Embryo Cells *in vitro, Proc. Soc. Exp. Biol. Med.* 121: 1260 (1966).

CELL-FREE PREPARATIONS

INDICATORS FOR PESTICIDES

A. Cholinesterase

Archer, T. E., W. L. Winterlin, G. Zweig, and H. F. Beckman. Residue Analysis of Ethion by Cholinesterase Inhibition after Oxidation, *J. Agr. Food Chem.* 11: 471 (1963).
Archer, T. E., and G. Zweig. Direct Colorimetric Analysis of Cholinesterase-Inhibiting Insecticides With Indophenyl Acetate, *J. Agr. Food Chem.* 7: 178 (1959).
Baum, G., and F. B. Ward. Ion-Selective Electrode Procedure for Organophosphate Pesticide Analysis, *Anal. Chem.* 43: 947 (1971).
Beynon, K. I., L. Davies, K. Elgar, and G. Stoydin. Analysis of Crops and Soils for Residues of Diethyl 1-(2,4-Dichlorophenyl)-2-Chlorovinyl Phosphate. I. Development of Method, *J. Sci. Food Agric.* 17: 162 (1966).
Beynon, K. I., and G. Stoydin. Application of an Agar-Agar Diffusion Procedure to Pesticide Residue Analysis and to the Cholinesterase Screening of Candidate Pesticides, *Nature (London)* 208: 748 (1965).
Bluman, N. Phosdrin Residues in Fruits and Vegetables, *J. Ass. Offic. Anal. Chem.* 47: 272 (1964).
Boyd, G. R. Determination of Residues of *o*-2,4-Dichlorophenyl *o,o*-Diethyl Phosphorothioate (V-C 13 Nemacide) by Cholinesterase Inhibition, *J. Agr. Food Chem.* 7: 615 (1959).
Bunyan, P. J. The Detection of Organo-phosphorous Pesticides on Thin-Layer Chromatograms, *Analyst* 89: 615 (1964).
Cook, J. W. Paper Chromatography of Some Organic Phosphate Insecticides. IV. Spot Test for *In Vitro* Cholinesterase Inhibitors, *J. Ass. Offic. Anal. Chem.* 38: 150 (1955).
Cook, J. W. Paper Chromatography of Some Organic Phosphate Insecticides. V. Conversion of Organic Phosphates to *In Vitro* Cholinesterase Inhibitors by N-Bromosuccinimide and Ultraviolet Light, *J. Ass. Offic. Anal. Chem.* 38: 826 (1955).
Crosby, D. G., E. Leitis, and W. L. Winterlin. Photodecomposition of Carbamate Insecticides, *J. Agr. Food Chem.* 13: 204 (1965).
Crossley, J. GLC and TLC Determination of Tetraethyl Pyrophosphate (TEPP) in Crops, *J. Ass. Offic. Anal. Chem.* 53: 1036 (1970).
Ebing, W. Über die Sprühreagens zum Dünnschichtchromatographischen Nachweis Cholinesterase hemmender Insektizide, *J. Chromatogr.* 42: 140 (1969).

El-Refai, A., and T. L. Hopkins. Thin-Layer Chromatography and Cholinesterase Detection of Several Phosphorothiono Insecticides and Their Oxygen Analogs, *J. Agr. Food Chem.* 13: 477 (1965).

Fallscheer, H. O., and J. W. Cook. Studies on the Conversion of Some Thionophosphates and a Dithiophosphate to *In Vitro* Cholinesterase Inhibitors, *J. Ass. Offic. Anal. Chem.* 39: 691 (1956).

Getz, M. E., and S. J. Friedman. Organophosphate Pesticide Residues: A Spot Test for Detecting Cholinesterase Inhibitors, *J. Ass. Offic. Anal. Chem.* 46: 707 (1963).

Giang, P. A., and S. A. Hall. Enzymatic Determination of Organic Phosphorous Insecticides, *Anal. Chem.* 23: 1830 (1951).

Guilbault, G. G., D. N. Kramer, and P. L. Cannon, Jr. Electrochemical Determination of Organophosphorous Compounds, *Anal. Chem.* 34: 1437 (1962).

Guilbault, G. G., S. S. Kuan, and M. H. Sadar. Purification and Properties of Cholinesterases from Honeybees — *Apis mellifera* Linnaeus — and Boll Weevils — *Anthonomus grandis* Boheman, *J. Agr. Food Chem.* 18: 692 (1970).

Guilbault, G. G., M. H. Sadar, S. Kuan, and D. Casey. Effect of Pesticides on Liver Cholinesterases from Rabbit, Pigeon, Chicken, Sheep, and Pig, *Anal. Chim. Acta* 51: 83 (1970).

Guilbault, G. G., M. H. Sadar, S. S. Kuan, and D. Casey. Enzymatic Methods of Analysis. Trace Analysis of Various Pesticides With Insect Cholinesterase, *Anal. Chim. Acta* 52: 75 (1970).

Gunther, F. A., and D. E. Ott. Automated Pesticide Residue Analysis and Screening, *Residue Rev.* 14: 12 (1966).

Kramer, D. N., and R. M. Gamson. Analysis of Toxic Phosphorous Compounds, *Anal. Chem.* 29(12): 21A (1957).

Lau, S. C. Separation and Measurement of 3-Hydroxy-*N,N*-dimethyl-*cis*-crotonamide Dimethyl Phosphate (Bidrin Insecticide) and 3-Hydroxy-*N*-methyl-*cis*-crotonamide Dimethyl Phosphate (Azodrin Insecticide) in Crops by Selective Cleanup (Partition) Procedures, *J. Agr. Food Chem.* 14: 145 (1966).

Leegwater, D. C., and H. W. Van Gend. Automated Differential Screening Method for Organophosphorous Pesticides, *J. Sci. Food Agric.* 19: 513 (1968).

Matoušek, J., J. Fischer, and J. Cerman. Nová Fluorimetrická Metoda Stanoveni Submikrogramových Kvant Inhibitoru Cholinesterázy, *Chem. Zvesti* 22: 184 (1968).

Mattson, A. M., R. A. Kahrs, and R. T. Murphy. Routine Quantitative Residue Determinations of S-[(2-Methoxy-5-oxo-Δ^2-1,3,4-thiadiazolin-4-yl) methyl] *O,O*-dimethyl phosphorodithioate (Supracide) and its Oxygen Analog in Forage Crops, *J. Agr. Food Chem.* 17: 565 (1969).

McCaulley, J., and J. W. Cook. The In Vitro Anticholinesterase Effect of Oxidized Parathion, Methyl Parathion, and Malathion on Eight Different Sources of Cholinesterase, *J. Ass. Offic. Anal. Chem.* 42: 197 (1969).

Menn, J. J., and J. B. McBain. Detection of Cholinesterase-Inhibiting Insecticide Chemicals and Pharmaceutical Alkaloids on Thin-Layer Chromatograms, *Nature (London)* 209: 1351 (1966).

Miskus, R., M. E. Tzanakakis, and S. M. Smith. Determination of Bayer 19639 Residues in Agricultural Crops by Cholinesterase Inhibition, *J. Econ. Entomol.* 52: 76 (1959).

Moorefield, H. H., and E. R. Tefft. Application of Cholinesterase Assay to Residue Analysis of 1-Naphthyl *N*-methylcarbamate (Sevin), *Contrib. Boyce Thompson Inst.* 19: 295 (1958).

Nesheim, E. D., and J. W. Cook. Cholinesterase Inhibition Method of Analysis for Organic Phosphate Pesticides: Effect of Enzyme-Inhibitor Reaction Time Upon Inhibition, *J. Ass. Offic. Anal. Chem.* 42: 187 (1959).

Ortloff, R., and P. Franz. Zwei Neue Methoden der Biochemischen Lokalisierung von Phosphorohaltigen Insektiziden auf Dünnschichtchromatogrammen, *Z. Chem.* 5: 388 (1965).

Ott, D. E. Dual Simultaneous Auto Analyzer for Screening Some Insecticide Residues, *J. Agr. Food Chem.* 16: 874 (1968).

Ott, D. E., and F. A. Gunther. Rapid Screening for Some Anticholinesterase Insecticide Residues by Automated Analysis, *J. Ass. Offic. Anal. Chem.* 49: 662 (1966).

Ott, D. E., and F. A. Gunther. Automated Elution-Filtration Analysis of Anticholinesterase Organophosphorous Compounds on Thin Layer Chromatographic Scrapings, *J. Ass. Offic. Anal. Chem.* 49: 669 (1966).

Ott, D. E., and F. A. Gunther. Procedure for the Analysis of Technical-Grade Parathion in Waterplants by an Anticholinesterase (AutoAnalyzer) Method, *J. Econ. Entomol.* 59: 227 (1966).

Patchett, G. G., and G. H. Batchelder. Determination of Trithion Crop Residues by Cholinesterase Inhibition Measurement, *J. Agr. Food Chem.* 8: 54 (1960).

Rosenthal, N. R. Two Modifications of the Colorimetric Procedure for Determination of Serum Cholinesterase. Application to Trithion and Phosdrin, *J. Ass. Offic. Anal. Chem.* 43: 737 (1960).

Sadar, M. H., S. S. Kuan, and G. G. Guilbault. Trace Analysis of Pesticides Using Cholinesterase from Human Serum, Rat Liver, Electric Eel, Bean Leaf Beetle, and White Fringe Beetle, *Anal. Chem.* 42: 1770 (1970).

Sándi, E., and J. Wight. An Agar-Diffusion Method for the Estimation of Organic Phosphate Insecticides, *Chem. Ind. (London)*, 1961: 1161.

Schutzmann, R. L. Note on Improved Spray Reagents for TLC Fluorogenic Detection of Cholinesterase Inhibitors, *J. Ass. Offic. Anal. Chem.* 53: 1056 (1970).

Schutzmann, R. L., and W. F. Barthel. Indoxyl Acetate Spray Reagent for

Fluorogenic Detection of Cholinesterase Inhibitors in Environmental Samples, *J. Ass. Offic. Anal. Chem.* 52: 151 (1969).

Voss, G. Automated Determination of Activity and Inhibition of Cholinesterase With Acetylthiocholine and Dithiobisnitrobenzoic Acid, *J. Econ. Entomol.* 59: 1288 (1966).

Voss, G. Peacock Plasma, A Useful Cholinesterase Source for Inhibition Residue Analysis of Insecticidal Carbamates, *Bull. Environ. Contam. Toxicol.* 3: 339 (1968).

Voss, G. The Fundamental Kinetics of Cholinesterase Reaction With Substrates and Inhibitors in an Automated, Continuous Flow System, *Residue Rev.* 23: 71 (1968).

Voss, G. Cholinesterase Inhibition Autoanalysis of Insecticidal Organophosphates and Carbamates, *J. Ass. Offic. Anal. Chem.* 52: 1027 (1969).

Winter, G. D. Automated Enzymatic Assay of Organic Phosphate Pesticide Residues, *Ann. N. Y. Acad. Sci.* 87: 875 (1960).

Winter, G. D., and A. Ferrari. Automatic Wet Chemical Analysis as Applied to Pesticide Residues, *Residue Rev.* 5: 139 (1964).

Winterlin, W., G. Walker, and H. Frank. Detection of Cholinesterase-Inhibiting Pesticides Following Separation on Thin-Layer Chromatograms, *J. Agr. Food Chem.* 16: 808 (1968).

Yip, G., and J. W. Cook. A Comparison of Four Cholinesterase Methods of Analysis for Organic Phosphate Pesticides, *J. Ass. Offic. Anal. Chem.* 42: 194 (1959).

Zweig, G., and T. E. Archer. Residue Determination of Sevin (1-Naphthyl *N*-Methylcarbamate) in Wine by Cholinesterase Inhibition and Paper Chromatography, *J. Agr. Food Chem.* 6: 910 (1958).

B. Other Esterases

Ackermann, H. Dünnschichtchromatographisch-Enzymatischer Nachweis Phosphorganischer Insektizide. Aktivierung Schwacher Esterasehemmer, *J. Chromatogr.* 36: 309 (1968).

Geike, F. Dünnschichtchromatographisch-Enzymatischer Nachweis und zum Wirkungsmechanismus von Chlorkohlenwasserstoff-Insektiziden, *J. Chromatogr.* 44: 95 (1969).

Geike, F. Dünnschichtchromatographisch-Enzymatischer Nachweis und zum Wirkungsmechanismus von Chlorkohlenwasserstoff-Insektiziden. II. Nachweis durch Hemmung von Trypsin, *J. Chromatogr.* 52: 447 (1970).

Geike, F. Dünnschichtchromatographisch-Enzymatischer Nachweis von Carbamaten. I. Nachweis Insektiziden Carbamate mit Rinderleber-Esterase, *J. Chromatogr.* 53: 269 (1970).

Geike, F. Dünnschichtchromatographisch-Enzymatischer und Gaschromatographischer Nachweis von 4,4'-Dichlorbenzophenon und seinen Abbauprodukten, *J. Chromatogr.* 54: 282 (1971).

Guilbault, G. G., and D. N. Kramer. Fluorometric Determination of Lipase, Acylase, Alpha- and Gamma-Chymotrypsin and Inhibitors of these Enzymes, *Anal. Chem.* 36: 409 (1964).

Guilbault, G. G., and M. H. Sadar. Fluorometric Determination of Pesticides, *Anal. Chem.* 41: 366 (1969).

Matsumura, F., T. A. Bratkowski, and K. C. Patil. DDT: Inhibition of an ATP-ase in the Rat Brain, *Bull. Environ. Contam. Toxicol.* 4: 262 (1969).

McKinley, W. P., and P. S. Johal. Esterase Inhibition Technique for Detection of Organophosphorous Pesticides. II Simplified Version for Routine Checking, *J. Ass. Offic. Anal. Chem.* 46: 840 (1963).

McKinley, W. P., and S. I. Read. Esterase Inhibition Technique for the Detection of Organophosphate Pesticides, *J. Ass. Offic. Anal. Chem.* 45: 467 (1962).

Mendoza, C. E., D. L. Grant, B. Braceland, and K. A. McCully. Evaluation of Esterases from Livers of Beef, Pig, Sheep, Monkey, and Chicken for Detection of Some Pesticides by Thin-Layer Chromatographic-Enzyme Inhibition Technique, *Analyst* 94: 805 (1969).

Mendoza, C. E., and J. B. Shields. Sensitivity of Pig Liver Esterase in Detecting Twelve Carbamate Pesticides on Thin-Layer Chromatograms, *J. Chromatogr.* 50: 92 (1970).

Mendoza, C. E., and J. B. Shields. Esterase Specificity and Sensitivity to Organophosphorous and Carbamate Pesticides: Factors Affecting Determination by Thin-Layer Chromatography, *J. Ass. Offic. Anal. Chem.* 54: 507 (1971).

Mendoza, C. E., and P. J. Wales. Liver Esterases of Rhesus Monkey: Inhibition and Activation by Selected Pesticides, *J. Agr. Food Chem.* 18: 503 (1970).

Mendoza, C. E., P. J. Wales, H. A. McLeod, and W. P. McKinley. Enzymatic Detection of Ten Organophosphorous Pesticides and Carbaryl on Thin-Layer Chromatograms: An Evaluation of Indoxyl, Substituted Indoxyl and 1-Naphthyl Acetates as Substrates of Esterases, *Analyst* 93: 34 (1968).

Mendoza, C. E., P. J. Wales, H. A. McLeod, and W. P. McKinley. Thin-Layer Chromatographic-Enzyme Inhibition Procedure to Screen for Organophosphorous Pesticides in Plant Extracts Without Elaborate Cleanup, *Analyst* 93: 173 (1968).

Ooms, A. J. J., and J. C. A. E. Breebaart-Hansen. The Reaction of Organophosphorous Compounds With Hydrolytic Enzymes. The Inhibition of Horse Liver Aliesterase, *Biochem. Pharmacol.* 14: 1727 (1965).

Ooms, A. J. J., J. C. A. E. Breebaart-Hansen, and B. I. Ceulen. The Reaction of Organophosphorous Compounds With Hydrolytic Enzymes. II. The Inhibition of Citrus Acetylesterase, *Biochem. Pharmacol.* 15: 17 (1966).

Ooms, A. J. J., and C. van Dijk. The Reaction of Organophosphorous Compounds With Hydrolytic Enzymes. III. The Inhibition of Chymotrypsin and Trypsin, *Biochem. Pharmacol.* 15: 1361 (1966).

Villeneuve, D. C., A. G. Butterfield, and K. A. McCully. A Carboxylesterase Inhibition Assay to Estimate Parathion, Malathion, and Diazinon in Lettuce Extracts, *Bull. Environ. Contam. Toxicol.* 4: 232 (1969).

Villeneuve, D. C., and W. P. McKinley. Inhibition of Beef Liver Hydrolytic Enzymes by Organophosphorous Pesticides, *J. Agr. Food Chem.* 16: 290 (1968).

Villeneuve, D. C., G. Mulkins, K. A. McCully, and W. P. McKinley. The Inhibition of Beef Liver Hydrolytic Enzymes by Organophosphorous Pesticides — A Comparison of the Effects of Several Pesticides and Their Oxons on the Inhibition Response, *Bull. Environ. Contam. Toxicol.* 4: 39 (1969).

Wales, P. J., H. A. McLeod, and W. P. McKinley. TLC-Enzyme Inhibition Procedure to Detect Some Carbamate Standards and Carbaryl in Food Extracts, *J. Ass. Offic. Anal. Chem.* 51: 1239 (1968).

Wales, P. J., C. E. Mendoza, H. A. McLeod, and W. P. McKinley. Procedure for Semiquantitative Confirmation of Some Organophosphorous Pesticide Residues in Plant Extracts, *Analyst* 93: 691 (1968).

C. Other Enzymes

Colvin, H. J., and A. T. Phillips. Inhibition of Electron Transport Enzymes and Cholinesterases by Endrin, *Bull. Environ. Contam. Toxicol.*, 3: 106 (1968).

Freedland, R. A., and L. Z. McFarland. The Effect of Various Pesticides on Purified Glutamate Dehydrogenase, *Life Sci.* 4: 1735 (1965).

Keller, H. Die Bestimmung kleinster Mengen DDT auf Enzymanalytischen Wege, *Naturwissenschaften* 39: 109 (1952).

Pardini, R. S. Polychlorinated Biphenyls (PCB): Effect on Mitochondrial Enzyme Systems, *Bull. Environ. Contam. Toxicol.* 6: 539 (1971).

Pardini, R. S., J. C. Heidker, and B. Payne. The Effect of Some Cyclodiene Pesticides, Benzenehexachloride and Toxaphene on Mitochondrial Electron Transport, *Bull. Environ. Contam. Toxicol.* 6: 436 (1971).

INDICATORS FOR CATIONS

Bamberger, C. E., J. Botbol, and R. L. Cabrini. Inhibition of Alkaline Phosphatase by Beryllium and Aluminum, *Arch. Biochem. Biophys.* 123: 195 (1968).

Guilbault, G. G., and D. N. Kramer. An Electrochemical Method for the Determination of Glucosidase and Mercuric Ion, *Anal. Biochem.* 18: 313 (1967).

Guilbault, G. G., D. N. Kramer, and P. L. Cannon, Jr. Electrochemical Determination of Xanthine Oxidase and Inhibitors, *Anal. Chem.* 36: 606 (1964).

Hughes, R. B., S. A. Katz, and S. E. Stubbins. Inhibition of Urease by Metal Ions, *Enzymolgia* 36: 332 (1969).

Kratochvil, B., S. L. Boyer, and G. P. Hicks. Effects of Metals on the Activation and Inhibition of Isocitric Dehydrogenase. Application to Trace Metal Analysis, *Anal. Chem.* 39: 45 (1967).

Mealor, D., and A. Townshend. Applications of Enzyme-Catalyzed Reactions in Trace Analysis. I. Determination of Mercury and Silver by Their Inhibition of Invertase, *Talanta* 15: 747 (1968).

Mealor, D., and A. Townshend. Applications on Enzyme-Catalyzed Reactions in Trace Analysis. III. Determination of Silver and Thiourea by Their Combined Inhibition of Invertase, *Talanta* 15: 1371 (1968).

Shaw, W. H. R., and D. N. Raval. The Inhibition of Urease by Metal Ions at pH 8.9, *J. Amer. Chem. Soc.* 83: 3184 (1961).

Thomas, M., and W. N. Aldridge. The Inhibition of Enzymes by Beryllium, *Biochem. J.* 98: 94 (1966).

Toren, E. C., Jr., and F. J. Burger. Trace Determination of Metal Ion Inhibitors of the Glucose-Glucose Oxidase System, *Mikrochim. Acta* 1968: 538.

Townshend, A., and A. Vaughan. Determination of Traces of Barium by Reactivation of Zinc-Inhibited Alkaline Phosphatase, *Anal. Lett.* 1: 907 (1968).

Townshend, A., and A. Vaughan. Applications of Enzyme-Catalyzed Reactions in Trace Analysis. IV. Determination of Beryllium and Zinc by Their Inhibition of Calf-Intestinal Alkaline Phosphatase, *Talanta* 16: 929 (1969).

Townshend, A., and A. Vaughan. Detection of Traces of Zinc by Activation of Apoalkaline Phosphatase, *Anal. Chim. Acta* 49: 366 (1970).

Townshend, A., and A. Vaughan. Applications of Enzyme-Catalyzed Reactions in Trace Analysis. V. Determination of Zinc and Calcium by Their Activation of the Apo-enzyme of Calf-Intestinal Alkaline Phosphatase, *Talanta* 17: 289 (1970).

INDICATORS FOR ANIONS

Cimasoni, G. Inhibition of Cholinesterases by Fluoride *in vitro, Biochem. J.* 99: 133 (1966).

Denburg, J., and W. D. McElroy. Anion Inhibition of Firefly Luciferase, *Arch. Biochem. Biophys.* 141: 668 (1970).

Kremer, M. L. Inhibition of Catalase by Cyanide, *Isr. J. Chem.* 8: 799 (1970).

Krupka, R. M. Fluoride Inhibition of Acetylcholinesterase, *Mol. Pharmacol.* 2: 558 (1966).

Mealor, D., and A. Townshend. Applications of Enzyme-Catalyzed Reactions in Trace Analysis. II. Determination of Cyanide, Sulphide, and Iodine With Invertase, *Talanta* 15: 1477 (1968).

Slater, E. C., and W. D. Bonner. The Effect of Fluoride on the Succinic Oxidase System, *Biochem. J.* 52: 185 (1952).

Yang, S. F., and G. W. Miller. Biochemical Studies of the Effect of Fluorides on Higher Plants. 2. The Effect of Fluoride on Sucrose-Synthesizing Enzymes from Higher Plants, *Biochem. J.* 88: 509 (1963).

Yurow, H. W., D. H. Rosenblatt, and J. Epstein. Detection of Monobasic Phosphorous Acid Esters by Conversion to Cholinesterase Inhibitors, *Talanta* 5: 199 (1960).

INDICATORS FOR AIR POLLUTANTS

Estes, F. L., and C. H. Pan. Response of Enzyme Systems to Photochemical Reaction Products, *Arch. Environ. Health* 10: 207 (1965).

Mudd, J. B. Responses of Enzyme Systems to Air Pollutants, *Arch. Environ. Health* 10: 201 (1965).

Ordin, L., M. J. Garber, J. I. Kindinger, S. A. Whitmore, L. C. Greve, and O. C. Taylor. Effect on Peroxyacetyl Nitrate (PAN) in vivo on Tobacco Leaf Polysaccharide Synthetic Pathway Enzymes, *Environ. Sci. Technol.* 5: 621 (1971).

Ordin, L., and M. A. Hall. Studies on Cellulose Synthesis by a Cell-Free Oat Coleoptile Enzyme System: Inactivation by Airborne Oxidants, *Plant Physiol.* 42: 205 (1967).

Ordin, L., M. A. Hall, and J. I. Kindinger. Oxidant-Induced Inhibition of Enzymes Involved in Cell Wall Polysaccharide Synthesis, *Arch. Environ. Health* 18: 623 (1969).

USE OF SENSE OF SMELL
IN DETERMINING ENVIRONMENTAL QUALITY

Trygg Engen

Professor of Psychology
Walter S. Hunter Laboratory of Psychology
Brown University
Providence, Rhode Island 02912

The avoidance of harmful situations and agents often depends on man's ability to assess odor characteristics. Although a few odors, for example, lemon, are considered pleasant and clean and are used for commercial purposes, odors are generally considered unhealthy annoyances to be removed from the air. This is in line with the view held by the World Health Organization's definition of health[26] as "a state of complete physical, mental, and social well-being and not merely the absence of disease or infirmity." The present emphasis will be the subjective or mental experience of odor. No attempt will be made to describe the physical or chemical attributes of the sources of odors; neither will any space be devoted to the physiological or anatomical characteristics of the olfactory system.[20]

ODOR DESCRIPTION

Classification

Most of the research on human characterization of odors has been directed toward classification of odors according to their perceptual similarities. Although there may have been some progress in developing a common glossary,[17] no agreement has been reached on a set of terms that might be as useful for communication about odors as the "primary colors" are in vision. Even if such agreement should be accomplished among investigators, the question is whether or not the public actually uses such terms when they describe or discriminate between odors. This seems unlikely. Then follows the basic question regarding the relationship between the verbal descriptions of odors and the physics of the odorant, and the physiology of the olfactory system.

Development of so-called multidimensional scaling techniques has made possible a more mathematically sophisticated approach to the odor classification problem.[10] These techniques do not use the subject's own labels for odors but simplify his task by asking him only to rate the similarity of the odors on some simple scale. From such judgments, it is possible to determine analytically the

133

number of ways in which the odors appear to differ subjectively, and one may thus obtain clues to the physical and physiological basis of these subjective differences. Although this approach is more sophisticated than the classic approach to odor classification, no substantial results have yet been produced. One fundamental problem may be that human observers may emphasize different attributes in different situations or for different sets of odorants. For example, a qualitatively diverse sample of odorants may suggest differences in pleasantness, whereas a more limited sample of the same odorants may seem to vary primarily in intensity. Therefore, it may not be possible to describe odors by a fixed number of dimensions. Like human motivation, the sense of smell may be a more dynamic system characterized by both contextual and temporal variations. In any case, such general principles of odor classification are not necessary for the practical problem of judging the quality of the environment.

Identification

This is a simpler problem than classification, and work on it has produced results that have not always been in accord with popular opinion. In recent unpublished experiments with large groups of college students, it was found that they were able to identify about 33% of a group of 20 common substances such as vinegar, lemon extract, menthol, etc. If one allows a more liberal criterion and gives credit for good associations related to the use of the substance (for example, calling menthol "Vicks Vapor"), the score increases to about 50%. The number of correct judgments increased to about 65% when the task was changed to recognition (as opposed to absolute identification) by providing the subject with a list of labels and instructing him to select one label for each odor.

These results agree well with results from analysis of odor identification judgments.[9] In general, it may be concluded that man is able to identify no more than about 16 different odors regardless of the number and characteristics of the odorants used to test him. This result is obtained when he is presented one odorant at a time and asked to identify it by the use of his sense of smell alone, without the help of any kind of comparison of odors or the use of any kind of non-olfactory cues. According to the Shannon-Wiener measure of information, this amounts to about 4 bits per stimulus.

Despite the fact that these results have been replicated several times, they have not found easy acceptance. It has long been believed, but not proved objectively, that the score should be much higher and that trained chemists and perfumers certainly could do much better. Jones' study[18] with such observers should dispel such a view, for he found that they did only a little better than our untrained academicians even though these experts had the advantage of using their own chemical compounds. (They also chose to remain anonymous!)

ODOR DETECTION

Although there is a limit to the ability of naive as well as trained observers to identify and recognize odors, the ability is potentially good enough for detection and identification of the relatively limited sources of odorous pollution. The most popular index of human sensitivity in olfaction as well as in other sense modalities has been the stimulus threshold. The nose has often been found to be quite sensitive as compared with available physical instruments. For example, only a few molecules of mercaptan appear to be necessary to stimulate the perception of an odor. In addition, threshold methodology is very easy to administer and requires only a simple judgment from the subject.

It is important to keep in mind that there are several kinds of threshold, some of which will be discussed at the end of this paper. The one to be considered most extensively here is the minimal detectable intensity of a substance, the so-called absolute threshold. This is the threshold considered in connection with setting standards for maximal allowable amounts of pollutants in the air. Another such index is the difference threshold which refers to the resolving capacity of the human observer. Given exposure to a certain amount of stimulation, by how much must that stimulation be changed physically before the observer becomes aware of it? While man may be very sensitive as measured by the absolute threshold, this ability to detect differences in the concentration of an odorant is usually poor. As measured by the difference threshold, the sense of smell has been shown to be one of the dullest sense modalities. For example, a person might become aware of a stimulus change of less than 1% in audition and vision; in the case of smell, the change required might be more than 25% before he will be able to detect that the smell is now stronger or weaker than it was.[24] This kind of comparison of the human nose with physical instruments would lead to a different conclusion than popular ones which emphasize absolute sensitivity of the nose.

In addition, there is a more serious and general problem with the detection task. As early as 1899, Slosson demonstrated that so-called hallucinations are easily observed in case of odor perception. Recent Swedish studies of the discomfort experienced by people living near pulp mills have also revealed that there is a strong tendency to report the presence of odor in the absence of the corresponding odorant.[5] The problem with the human observer as an instrument is that he is biased toward reporting what he anticipates as well as what he actually experiences. For example, he may be more likely to report experiencing odors when he sees smoke. In order to apply human detection judgments as indicators of environmental quality, it is necessary to devise methods that make it possible to measure odor sensitivity independently of the observer's attitude toward the source of the odorant.

To define this problem more clearly, one simple experiment[11] will be discussed in more detail. The experiment was designed to assess the effect of variables that are unrelated to the odorant but which may affect the human observer's expectation and thus his report of what he perceives. Three factors were varied: the (irrelevant) color of the odorant, the "feedback" information provided the subject about the accuracy of his judgments, and the probability that an odorant versus an odorless "blank" would be presented.

The first part of the study was concerned with the irrelevant color of the odorant: a 10-millimole concentration of butanol in ethyl phthalate was presented to the subject on a cotton swab. Half of the cotton swabs presented to him for sniffing had been tinted a desaturated yellow with an odorless dye (Sudan Yellow), and the other half contained only the transparent liquid, labeled "clear" for the present discussion. On any one trial the subject might receive any one of four different swabs: a clear blank (only the diluent), a tinted blank, a clear odorant (butanol diluted in ethyl phthalate), or a tinted odorant. The probability of the presentation of each event was 0.25. Before the experiment started, the subject was presented both an odorant and a blank in order to familiarize him with both, and it was explained that his task would be to judge whether one or the other had been presented on a particular trial. Ethyl phthalate is not judged odorless by all people, but the subject was instructed to respond to it with "no." None of the subjects had any problem whatever distinguishing between the odorant and the blank, although smelled by itself, singly, the concentration of butanol is weak and not perfectly detectable. As mentioned above, comparison of stimuli is apt to be a much easier task than identification of single stimuli.

After the preliminary part, the procedure was presented according to a simple Yes-No detection experiment shown below. On each trial either an odorant (clear or tinted) or a blank (clear or tinted) was presented and the observer was instructed to respond either "yes" or "no," with the possible outcome shown in the following stimulus-condition matrix:

Response	Odorant	Blank
Yes	Hit	False alarm
No	Miss	Correct rejection

Since the observer had to respond either "yes" or "no," the percentage of hits plus misses and false alarms plus correct rejections will add to 100 in each case; therefore only the percentage of hits and false alarms need be presented. The whole experiment required 4 days of testing per subject. Although the results for only one subject will be presented here, the experiment was repeated with two other subjects with the same results.

The following results were obtained for the first experiment on clear versus tinted odorants. Each percentage is based on 500 trials.

	Clear odorant	Tinted odorant
Hit	74%	74%
False alarm	60%	69%

The most interesting aspects of the present data are that (1) the percentage of false alarms is very high for both conditions and (2) it is higher for tinted odorants than clear odorants. Despite the preliminary work on the definition of "no odor" and the practice trials regarding the difference between the odorant and the blank, all subjects show this tendency to mistake a blank for an odorant. It is clear that the criterion for reporting the experience of an odor is very low. By comparison, such high percentages of false alarms are unlikely in audition and vision.

Since it is often considered that the sense of smell suffers from disuse in modern society, the next part of the experiment involved the use of feedback information about the accuracy of the subject's reports about the stimuli. For every correct judgment (hit or correct rejection), the experimenter said "correct," and in addition the subject was rewarded by an extra penny for that trial. For every incorrect judgment, false alarm, or miss, the experimenter said "wrong" and the subject was fined one penny. The losses and gains were recorded by the subject on counters, and on the average this payoff amounted to a substantial increment of $1.00 to the subject's hourly rate of $1.60.

The data obtained for the same subject are shown in the following matrix:

	Clear odorant	Tinted odorant
Hit	34%	53%
False alarm	20%	31%

The payoff clearly made the subject much less likely to say "yes" or, in other words, to report the experience of odor regardless of the nature of the trial. Yet the biased effect of the visual cue is now even more noticeable than before, because both the percentages of hits and false alarms are greater for tinted than clear swabs. It had been expected that the opposite would happen; that is, irrelevancy of the tint should become evident with further experience and feedback information. Such results can probably be obtained with still further training, but the bias is apparently strong enough to resist extinction even after 500 feedback trials. The subjects apparently were not aware of this bias in their judgments, because when questioned about their opinion regarding the tinted versus clear odorant at the completion of the experiment, they all indicated individually that the difference in color had been noticed but that they had found it to be irrelevant!

The third part of this experiment studied the effect of the expectation of the subject that an odor would be perceived by varying the odorant presentation probability. In the experiments described above, this probability was 0.50; that is, odorants were presented in half of the trials and blanks in the other half. In the last part of the experiment, the subjects were tested under odorant presentation probabilities of 0.10 and 0.90. In other words, under the former condition the odorant was presented in 10% and the blank in 90% of the trials, and for the latter condition, the reverse was the case. Only clear odorants were used in this part of the study. The results (including the results for 0.50 from above) were:

	Odorant presentation probability		
	0.10	0.50	0.90
Hit	32%	34%	97%
False alarm	4%	20%	78%

It is clear that when the probability of the occurrence of an odorant is increased, the tendency to report the presence of an odor also increases. However, this increase may not be taken as evidence of the veridicality of perception, as one might by considering percentages of hit only, because increase in the presentation probability also increases the percentage of false alarms. In other words, the main effect of increased frequency of odorant presentation may be on the expectation of the subject, and for that reason he tends to respond "yes" more often both to the odorant and to the blank.

The main conclusion to be drawn from these findings is that the classical threshold or "frequency of smelling" curve does not provide an adequate basis for the assessment of the quality of the air. A higher order of assessment that takes into account the response tendencies of the observer is necessary. Modern mathematical detection models the result of the combined efforts of psychologists and engineers and does provide the basis for such an approach.[16]

In general, this theory assumes that the experience of an odor is possible at any moment because of the inherent variability of the sensory system and the environment. However, the percentage of false alarms may be used as an estimate of the amount of "noise" in the perceptual system, that is, the magnitude of odor experience unrelated to the experimental odorant. The effect of the odorant is added to the amount associated with noise along the underlying perceptual dimension used by the observer in forming his judgment. His attitude or bias is thought to affect the criterion for the perceived magnitude used by the observer to divide this dimension into two categories as required by the experiment – those experiences described by "yes, I smell the odor" versus "no, I do not smell anything." The lower his criterion, the more likely he is to respond "yes" to a certain weak odor.

The variables affecting the observer's criterion, for example, his attitude, are independent of the variables affecting the perceptual magnitudes. As a result, a change in the criterion will affect the functional relationship between the proportion of false alarms and proportion of hits but not the extent to which an added signal displaces the noise distribution. This makes it possible to estimate the signal strength by comparing the proportion of false alarms, as an estimate of the noise, with the proportion of hits as the effect of signal-plus-noise. What is needed to develop the best model for practical application is quantitative data on the mathematical relationship between false alarms and hits for different odorants, environmental contexts, and observers in various states of adaptation.

Application of such an analysis to the present data indicates that the effect of the odorant is much more constant than the varying proportions of false alarms and hits might lead one to believe. The development of such a mathematically sophisticated model of measurement for olfaction is both necessary and possible judging by the research already done.[2,6,22] However, one should not lose sight of the fact that the study of detection is only part of a more general approach to the odor problem.

ODOR SCALING

It has generally been assumed that the simplest quantitative response one can obtain from a human subject is the best one and that the more demanding the numerical task of the subject, the less reliable and "scientific" the results. Methods for measuring threshold have been considered ideal in requiring only a response of the "presence" or "absence" of an odor and yet providing basic information about the transduction capacity of a sense organ. The obvious shortcoming of the threshold and similar indices of detection is that neither qualitative nor hedonic differences in stimulation are elicited at this level of detection. Neither is there any physiological knowledge regarding receptors and coding to help the experimenter predict what kind of quality his subject might best discriminate. A generally useful science of smell demands consideration of both threshold and suprathreshold levels of stimulation. One should like to know, for example, whether or not man is able to sense that the concentration of some odorous and dangerous pollutants has increased to such a degree that he should leave the scene. Fortunately, there is already what might rightly be called the rudiments of such a science of smell. Some small inroads have been made in the territory often assumed to be both unscientific and virtuous and the sole prerogative of the artist.

Not only does the odor problem demand study of suprathreshold odors, but a case can be made for the advantages of suprathreshold methods on purely technical grounds. Although most scientists find this difficult to believe, suprathreshold judgments are apparently more orderly and easier to obtain than threshold data. Threshold represents by definition a boundary where stimulus control fails and the response becomes unreliable; it refers to a situation in which the subject is not sure whether or not he perceives the odorant. That is one reason for the difference in results usually obtained from different methods. It is really not surprising that an answer to the question by Swets[25] "Is there a sensory threshold?" is still in doubt 100 years after Fechner[14] proposed the methods for measuring it.

The measurements made by scientifically inclined students of olfaction are usually not intrinsically psychological. In order to avoid what is considered the subjectivity of perception, physical variables associated with it, such as vapor pressure and concentration in the case of olfaction, have dominated the field.

There is one basic reason why such physical (or chemical) measurements are not sufficient, despite their obviously essential role, and why a more direct psychological approach is necessary. Even when a detection idex has been determined and, as is customary, statistically specified on some physical dimensions, one does not know anything about the effectiveness of other values along that dimension – even those very close to it. For example, when concentration exceeds the so-called absolute threshold, equal increments in terms of concentration usually will not be perceived as equal by the subject. Psychological scaling refers to the methods whereby one deals with that problem, that is, how perceived intensity and quality of odor vary as a function of physical variables.

The Effect of Intensity

The logarithmic psychophysical law Fechner proposed on the basis of Weber's work on difference threshold was long considered essentially correct and valid enough for practical application. However, during the last two decades, other so-called direct psychophysical scaling methods have been introduced. As a result, Fechner's law, according to which sensation grows linearly as a function of the logarithm of physical intensity, has been replaced by a power function, or Stevens's law, according to which sensation grows as a function of stimulus intensity raised to some power.[2,3]

The evidence obtained in scaling odor intensity also supports this power function,

$$R = k(S - S_0)^n, \tag{1}$$

where R is perceived or subjective intensity, S is physical intensity or concentration, S_0 is an estimate of threshold, k is a constant depending upon the units of measurement employed, and n is the power, that is, the parameter indicating the rate of growth of perceived intensity with increase in physical concentration. In olfaction, the parameter n, the exponent of the function, depends on the odorant and has been found to vary[1,3] from approximately 0.15 to 0.75.

There are also, of course, individual differences. However, a representative value of the exponent is about 0.6. Some of the results obtained may be described with other mathematical functions because the results are variable, but none describe all the data obtained in over a dozen different investigations on three dozen odorants better than the power function.[1]

For practical purposes of odor control, this mathematical function means that equal physical ratios of concentration correspond to equal subjective ratios of odor intensities. For example, if the concentration is doubled, this will only result in an increase of 1.7 times the subjective intensity for an exponent of about 0.6. If the concentration is halved, perceived intensity is reduced only 60%. The important point is that the relation between ratios holds for all detectable concentrations of the odorant. As noted, the value of the ratios may be different for different odorants. Since ratios are involved on both the psychological and the physical side of the equation, it follows that in plotting the perceived odor intensity against physical concentration, a straight line will be obtained when both (not only concentration, as proposed by Fechner) are converted to logarithms.

Methodologically, this kind of information is obtained by asking observers to describe on a ratio scale the perceived (or subjective) intensity of each of a range of concentrations by a number relating it to the perceived intensity of a standard. For example, the instructions might ask the observer: "If the subjective intensity of the odor from this source is described by 10, what number describes the intensity of the odor from the other source? If it smells twice as strong call it 20; if half as strong, 5; etc." When the average logarithmic value of the numerical judgments is plotted against the logarithmic value of each of the concentrations, a linear psychophysical function is usually obtained. This methodology and the power function have been supported by research in all the sense modalities. It may be concluded that this kind of suprathreshold psychophysics of odor intensity is advanced enough to be as useful for work on air quality as perceived loudness of sound intensity is in the case of auditory noise.

The Effect of Adaptation

The adaptive effect of one stimulus upon a subsequent stimulus of the same compound (self-adaptation) affects all the parameters of Equation (1). An increase in the concentration of an adapting stimulus presented before each of the stimuli judged by the observer tends to increase both the slope of the psychophysical function (n) and the threshold of detection (S_0) and to decrease k. It is particularly noticeable that the function becomes steeper for the lower stimulus concentrations and tends to reach an asymptote at the concentration corresponding to S_0 in Equation (1) above. In other words, the effectiveness of adaptation is inversely proportional to the concentration of the odorant judged such that the more adapted the observer, the higher his threshold and the greater the perceptual change in intensity for any given physical change in concentration. The effect of cross-adaptation, that is, prior exposure to one odorant on the perceived intensity of a qualitatively different odorant, also conforms to this rule.[4]

The effect of duration of adaptation, or the number of inhalations of the same adapting stimulus, is also similar, but, surprisingly, it is small compared with the effect of varying the intensity of the adapting stimulus. Ekman et al.[7] have proposed that this effect may be described as a simple exponential function,

$$R = \frac{a + b}{c^T}, \tag{2}$$

where R is perceived odor intensity, T duration, c the rate of the adaptation process, a the asymptote of the function, and a + b the value of R at T = O. Although threshold concentration (S_0) will increase as a function of duration of adaption, it reaches a new asymptotic level, and thus ability to detect odor rarely disappears completely even after prolonged exposure.[2] Research with modern detection methodology does not therefore verify Zwaardemaker's data[27] suggesting that sensitivity as measured by threshold decreases linearly with duration of exposure such that an observer would become insensitive after only several minutes of exposure. Instead, sensitivity seems to be relatively stable, although it depends on the adapting concentration. A reasonable conclusion is that the combined effect of concentration and duration of an adapting stimulus is to set the lower limit for sensitivity (S_0) of the olfactory system according to a signal-to-noise system and to increase the rate of growth (n) of perceived intensity as a function of stimulus intensity. Except for extreme and potentially harmful levels, adaptation does not bring the system to subjective zero or exhaustion but only affects the likelihood that an odorant or the difference in concentration of two odorants will be detected.

The Effect of Quality

It has often been observed that besides intensity the quality of some odors also changes with dilution, and this change may cause an orderly deviation from the power function. Some of our experiments have suggested that when a relatively naive subject is asked to estimate the subjective intensity of an odor, he may be unable to follow the experimenter's instructions to ignore such qualitative differences and may assign numbers which are proportional to the distinctiveness of the quality of the odor instead of its intensity. Zwaardemaker[28] commented on this problem that "...observers are generally not accustomed to distinguishing accurately the qualities and intensities of the smells they meet. ...Some discover chemical resemblances; others are strongly impressed by the agreeable or disagreeable effects connected with the sensations." The relationship between quality and intensity may be the reason that for some odorants the psychophysical function seems to be steeper for higher concentrations than for lower concentrations when, presumably, quality is becoming less distinct but the odor is still detectable. If the subjective zero for the quality of the odor is higher than the lowest concentrations presented, then one might make a correction for this by adding an amount on the response dimension corresponding to the response threshold for perceived quality and thus straighten the function in log-log coordinates. The function then becomes,

$$R = ks^n + R_0 , \qquad\qquad (3)$$

where R_0 is the response threshold for quality. This parameter (R_0) is potentially important and might correspond to a largely neglected "recognition threshold." Presumably, this threshold corresponds to the concentration at which n-heptane, for example, begins to smell like "gasoline." There is clearly a much lower concentration at which he can detect an odor but is unable to identify it in terms of such associations, and this performance is usually described[8] in terms of the "detection threshold."

The Effect of Unpleasantness

The relationship between changes in quality and acceptability of odors is not clear, for these attributes may also change concomitantly. There is a very strong suggestion, however, that observers tend to judge unpleasant odors to be strong; perhaps that is the common meaning of "pungent." For example, heptanal may be illustrative of this, for it is unpleasant to most people at all the detectable concentrations and difficult to judge solely on the dimension of intensity. For other odorants, preferences tend to vary substantially with intensity or concentration. For example, butyric acid may be attractive at low concentrations but very unpleasant at high concentrations. There is clearly a need for research on this relationship between pleasantness and perceived intensity. On the basis of present information, one would expect the correlation between them to be negative, but more refined observations need to be made.[15]

As indicated above, an index of detectability is important both for intensity and quality in scaling suprathreshold odors, and such an index can be specified quantitatively for both perceptual attributes. Is there a similarly manageable discomfort index? There is evidence that neutral and weak odors tend to be judged as pleasant, in agreement with the hypothesis suggested above and with Moncrieff's contention[19] that "he smells best who smells nothing." Perhaps because of the effect of the context provided by the odorants used in a particular experiment, it may be impossible to define a neutral odor to be used for experimental control conditions. On the other hand, there is some reason to suspect that affective judgments of odors may be more stable than those for other perceptual attributes.[21]

Psychophysical scaling methods have already been applied with success even in the case of stimuli for which the physical correlates are unknown and most likely very complex. In one study, the observers were asked to judge the pleasantness of a diverse sample of odorants with the method of magnitude estimation. Comparison of those results with judgments of intensity and pleasantness makes it very clear that there is a much greater variation in response associated with the latter attribute. The range of subjective values (numerical magnitude estimates) for the pleasantness may be more than 10 times the subjective range of intensity values corresponding to the whole dynamic range of concentration for a typical odorant.[12]

It must be borne in mind that while psychophysics can now describe these various perceived attributes at a fairly high level of quantification, the explanation of the hedonic value of an odor will require additional methods of analysis of psychological meaning. Whether or not a person likes or dislikes a certain odor depends largely upon his experience, the effect of cultural norms, and his present condition.[10]

A recent study[13] of odor preference of children and adults for odors, some of which undoubtedly had marked trigeminal effects (e.g., ammonia), showed clearly that children are more tolerant of odors than adults. Each child had to choose which member of a pair of odorants he preferred. Although the children were able to discriminate between the odorants, they were neither attracted nor repelled as much by them as were the adult subjects tested for comparison. For example, the children chose butyric acid as the preferred member of a pair when the adults would almost always choose the other as more pleasant. However, the older the child, the more likely it is that his choice will be similar to the adults. The younger the child, the more tolerant he seems. Although an individual child may express strong dislikes to an odor, the variability of children's preferences is greater than that of the preferences of adults. In other words, while children have a tendency to say they like odors, adults tend to say they dislike them, especially if the odors are unfamiliar.

There is, then, more to man's response to his environment than can be measured by physical and psychophysical instruments. The problem of the development of hedonic judgment brings us to the psychology of how response biases, preferences, and aversions have been learned. This too is very much a problem of environmental control but beyond the scope of the present symposium.

REFERENCES

1. Berglund, B., U. Berglund, G. Ekman, and T. Engen. Individual psychophysical functions for 28 odorants, *Perception and Psychophysics,* 9(3B): 379–384 (1971).
2. Berglund, B., U. Berglund, T. Engen, and T. Lindvall. The effect of adaptation on odor detection, *Perception and Psychophysics,* 9(5): 435–438 (1971).
3. Cain, W. S. Odor intensity: Differences in the exponent of the psychophysical function, *Perception and Psychophysics,* 6A: 349–354 (1969).
4. Cain, W. S., and T. Engen. Olfactory adaptation and the scaling of odor intensity, pp. 127–141 in *Olfaction and Taste III,* C. Pfaffmann (ed.), Rockefeller University Press, New York, 1969.
5. Cederlöf, R., L. Friberg, E. Jonsson, L. Kaij, and T. Lindvall. Studies of annoyance connected with offensive smell from sulfate cellulose factory, *Nordisk Hygienisk Tidskrift,* 45: 39–48 (1964).
6. Corbit, T. E., and T. Engen. Facilitation of olfactory detection, *Perception and Psychophysics,* 10: 433–436 (1971).
7. Ekman, G., B. Berglund, U. Berglund, and T. Lindvall. Perceived intensity of odor as a function of time of adaptation, *Scandinavian Journal of Psychology,* 8: 177–186 (1967).
8. Engen, T. Effect of practice and instruction on olfactory threshold, *Perceptual and Motor Skills,* 10: 195–198 (1960).

9. Engen, T. Man's ability to perceive odors, pp. 361–384 in *Advances in Chemoreception*, J. W. Johnston, Jr., D. G. Moulton, and A. Turk (eds.), Vol. 1. Appleton-Century-Crofts, New York, 1970.

10. Engen, T. Olfactory psychophysics, pp. 216–244 in *Handbook of Sensory Physiology*, L. M. Bedler (ed.)., Vol 14. Springer-Verlag, New York, 1971.

11. Engen, T. Method and theory in the study of odor preferences, in *Advances in Chemoreception*, Vol. II. *Human Responses to Environmental Odors*, J. W. Johnston, Jr., D. G. Moulton, and A. Turk (eds.). Appleton-Century-Crofts, New York in press.

12. Engen, T., and D. H. McBurney. Magnitude and category scales of the pleasantness of odors, *Journal of Experimental Psychology*, 68: 435–440 (1964).

13. Engen, T., and L. Moskowitz. The influence of the trigeminal stimulation on children's judgments of odor. Injury Control Research Laboratory, Publication No. ICRL-RR-71-4. U. S. Department of Health, Education, and Welfare, in press.

14. Fechner, G. T. *Elemente der Psychophysik*. Breitkopf and Harterl, Leipzig, 1860; English translation of Vol. 1 by H. E. Adler (D. H. Howe and E. G. Boring, eds.). Holt, Rinehart and Winston, New York, 1966.

15. Foster, D. Odors in series and parallel. Proceedings of Scientific Section, The Toilet Goods Association, 39: 1–6 (1963).

16. Green, D. M., and J. A. Swets. *Signal Detection Theory and Psychophysics*. John Wiley & Sons, New York, 1966.

17. Harper, R., E. C. Bate-Smith, and D. G. Land. *Odour Description and Odour Classification*. J & A Churchill Ltd., London, 1968.

18. Jones, F. N. Information content of olfactory quality, in *Theories of Odors and Odor Measurement*, N. Tanyolac (ed.). Robert College Research Center, Bebek, Turkey. Bantam Books, New York, 1968.

19. Moncrieff, R. W. *Odour Preferences*. John Wiley & Sons, New York, 1966.

20. Mozell, M. M. The chemical senses, II. Olfaction, pp. 193–222 in *Woodworth & Schlosberg's Experimental Psychology* (3rd Ed.), J. W. Kling and L. A. Riggs (eds.). Holt, Rinehart and Winston, Inc., New York, 1971.

21. Sandusky, A., and A. Porducci. Pleasantness of odors as a function of the immediate stimulus context, *Psychonomic Science*, 3: 321–322 (1965).

22. Semb, G. The detectability of the odor of butanol, *Perception and Psychophysics*, 4: 335–340 (1968).

23. Stevens, S. S. On the psychophysical law, *Psychological Review*, 64: 153–181 (1957).

24. Stone, H., and J. J. Bosley. Olfactory discrimination and Weber's law, *Perceptual and Motor Skills*, 20: 657–665 (1965).

25. Swets, J. A. Is there a sensory threshold?, *Science*, 134: 168–177 (1961).

26. World Health Organization, p. 1080 in *Yearbook of International Organizations*. Union of International Associations, Brussels, 1968–69.

27. Zwaardemaker, H. *L'Odorat*. Librairie Octave Doin, Paris, 1925.

28. Zwaardemaker, H. An intellectual history of a physiologist with psycho-
    logical aspirations, pp. 491–516 in *A History of Psychology in Auto-
    biography*, C. Murchison (ed.). Clark University Press, Worcester, Mass.,
    1930.

DEVELOPMENT OF ENVIRONMENTAL INDICES: OUTDOOR RECREATIONAL RESOURCES AND LAND USE SHIFT*

Robert P. Pikul

Head, Environmental Quality Sub-Department
MITRE Corporation
McLean, Virginia 22101

Charles A. Bisselle and Martha Lilienthal

Environmental Quality Sub-Department
MITRE Corporation
McLean, Virginia 22101

INTRODUCTION

Analog Between Economic and Environmental Indices

Economic and social indicators and indices such as the index of wholesale prices, index of retail sales, and the cost of living index have been utilized by governmental decision makers and the general public for years. More recently, environmentally related indices have been developed to assess the quality of our air, the quality of our water, and the environment in general.[2,4,12] These environmental indices serve a useful purpose, but they are somewhat limited in scope, failing to consider many important factors. The EQ (Environmental Quality) Index, published annually by the National Wildlife Federation, is strikingly illustrated and contains a variety of interesting, often alarming facts, but the Federation does not claim to be analytically rigorous in its derivation of quantitative results.

Because of the broad range of environmental factors that must be addressed, a single index, no matter how desirable from a standpoint of reporting simplicity, would not be adequate to highlight significant trends for each environmental area.

To illustrate this point, let us consider various economic indices. The gross national product, which is the total national output of goods and services valued at market prices, is regarded by many as the best single indicator of the state of

---

*The material presented in this paper is based upon a study performed by The MITRE Corporation for the Council on Environmental Quality related to a system concept for monitoring the environment of the nation.[8]

the economy. The GNP, reported since 1929, includes (1) private and federal purchase of goods and services, (2) gross private domestic investment, and (3) net exports of goods and services.[15] Even though the GNP is a useful index, it is informative to examine other economic indices such as those for wholesale prices, retail prices, unemployment, and cost of living. Such supplementary indices will always be useful, especially since many experts disagree as to the meaning and relative importance of the various indices.

In contrast to the highly developed economic indices, environmental indices[8,16] are just now being defined. Current knowledge about the environment is so rudimentary that identification and measurement of parameters for specific indices and indicators are extremely difficult. It would be premature to attempt to develop an overall index for the environment in any rigorous fashion, although subjective judgments of the "state of the environment" may be useful. Consequently, the first step in the development of environmental indices must be to focus upon specific areas of concern.

Illustrations have been borrowed from the field of economics for two primary purposes:

1. Economic indices and indicators, having been utilized and reported for many years, are fairly familiar to decision-makers and provide a good analogy to clarify basic concepts which will be described in subsequent sections of this paper.
2. Economic indices, in spite of their refined state of development and use relative to environmental indices, contain many conceptual difficulties such as:

   (a) What is a "typical" family and how does one measure the distribution of its expenditures for essential goods and services?
   (b) How does one determine the components of the basic commodity group and the essential goods and services to be included in the cost of living index?
   (c) What is the rationale for assigning weights to terms in an index?
   (d) How does one obtain representative samples?

These difficulties have been resolved on a practical basis, and in spite of many other problems related to definition, interpretation, and value judgments, the employment of economic indices to assist other processes of assessment, evaluation, and control has proven to be useful. Similar difficulties will be encountered with environmental indices, which emphasizes the necessity for continuing development and refinement.

Grouping of Environmental Indices

Environmental indices may be grouped as they relate to:

1. The natural and polluted environment
   (a) Emissions, (b) Quality, (c) Episodes

2. Effects

   (a) Human health, (b) Biosphere, (c) Materials and related economics

3. Resources

The following sections describe indices or indicators* of recreational resources and land use resources which are specific examples of the third grouping. The presentation provides a point of departure for further development and refinement which may be undertaken once data are collected and operational experience in measurement and use has been obtained.

INDICES OF OUTDOOR RECREATIONAL RESOURCES

Characterization of outdoor recreation areas in a systematic way is difficult because of factors such as the multiplicity of recreational and non-recreational uses of a given area, its intensity of use (e.g., the number of visitors per acre), and the wide variance associated with recreation land distribution. For example, in *Outdoor Recreation for America, A Report to the President,* the Outdoor Recreation Resources Review Commission in 1962 indicated that recreation areas of over 100,000 acres in size constitute about 1% of the number of areas but account for 88% of the total acreage. Acreage ranges in size from municipal or school parks of less than one acre to the 1.9-million-acre Mt. McKinley park in Alaska. Such disparity introduces difficulties in aggregating meaningful data on a national level. The importance of the nation's outdoor recreational resources is underscored by the growth and pressures placed on their utilization by an expanding population. Availability of leisure time, increased disposable income, and greater mobility also contribute to growth in utilization of recreation resources.

The Bureau of Outdoor Recreation compiled an inventory of approximately 20,000 public recreation areas in 1965. Data were provided by 12 federal bureaus and services in the Departments of Interior, Agriculture, and Defense, as well as by state, county, and other local agencies. While the inventory of public areas was reasonably complete (every city of population in excess of 50,000 was covered; a 10% sample of cities with population less than 50,000 was taken), much less data are available for private areas. This effort should be extended to provide annual updates of both changes in numbers of areas and, more importantly, changes in recreational facilities and activities associated with existing areas.

Information on the status of recreational resources is required for three distinct categories that are distinguished generally by the level of detail which is provided. The categories, described in more detail in subsequent paragraphs, are:

1. Gross characteristics of recreational areas
2. Recreational resource units for recreational areas
3. Recreational resource characteristics of major urban centers

---

*In the context of this discussion, an index is considered to be a mathematical combination of two or more parameters which has utility in an interpretive sense. An indicator calls attention to a potential change in a related phenomenon.

Parameter Requirements

The data required for each recreation area are shown in Table 1. In addition to the parameters shown, location data (e.g., latitude, longitude, county, state), type of agency, terrain features, and other indicative data contained in BOR Form 8–73 (used by the Bureau of Outdoor Recreation in compiling the Inventory of Public Outdoor Recreation Areas) should be obtained.

The reporting of usage should be standardized for purposes of computing indices to report either visits or visitor days. The latter measure is a better indication of total use since it takes into account overnight visits, but it may be more difficult to estimate. Since a large percentage of visits to recreational areas are 1 day or less, this development will consider visits rather than visitor days.

Recreation Availability Indices

Availability indicators should be based upon acreage and usage for acres within the National Park System[9] (Table 2) as well as for state parks. Basic trends in sizes of specific kinds of recreation areas such as urban parks, shorelines, and beaches should also be reported. Table 3 presents an aggregate list of types of recreation resources for which availability indices should be reported. It is suggested that visits be included only for state and national areas, while size (in terms of area or linear distance, whichever is appropriate) be included for all aggregate recreation area types. Indices will be based on the following specific reported measures:

S_{ij} = area of recreation site j for general type i. For example, i=1 could refer to national parks and j=1 could refer to a specific national park such as Yellowstone.

V_{ijk} = number of visits to site j for general type i in quarter k. k=1 could refer to January–March.

L_{ij} = linear measure of shore line and beaches for beach or lakesite j for beach or lake category i (national, state or other). Likewise for trails.

n_i = number of sites in general type i.

Spatial and Linear Availability. The areal space for each general type of recreation area, i, is obtained by summing S_{ij} for all n_i sites. Hence,

$$S_i = \sum_{j=1}^{n_i} S_{ij} . \tag{1}$$

Total recreational area is obtained by accumulating S_i over all recreation area types:

$$S = \sum_{i=1}^{n} S_i \quad (n = \text{total number of types listed in Table 1}). \tag{2}$$

Table 1. Data monitoring requirements for outdoor recreation

Parameter	Unit of measurement	Geographical extent	Method of collection	Frequency of collection	Potential source	Additional extension
Type of area	Park (national, state) Rec. area Beach National forest Other forest Hunting Fishing Wildlife Historic site Wayside Parking or recreation road Trail Waterway Nature preserve Other	County in state for each area	Survey form – cooperative reporting	3-yr Inventory, with annual update	1. Bureau of Outdoor Recreation 2. Dept. of Interior 3. Dept. of Agriculture 4. Dept. of Defense 5. Corps of Engineers 6. State and local sources	Inventory of public areas should be updated and revisions compiled annually in a systematic fashion. A good compilation for private areas is lacking.
Type of activity	Driving & sightseeing Swimming Water skiing Skin & scuba diving Bicycling Fishing Picnicking Nature study Boating Hunting Horseback riding Golf Camping Ice skating Sledding Snow skiing	Same as above	Same as above	Same as above	Same as above	Same as above

Table 1. (continued)

Parameter	Unit of measurement	Geographical extent	Method of collection	Frequency of collection	Potential source	Additional extension
Type of activity	Hiking Mountain climbing Playing games Viewing sports Dramatic & concert Other					
Physical size	Acreage of area (include miles of trails, shoreline, etc.)	Same as type of area	Same as type of area	Same as type of area	Same as type of area	Same as type of area
Capacity	No. of people (or visits), by quarter	Same as type of area	Same as type of area	Same as type of area	Same as type of area	Should be broken down by quarter to allow reporting seasonal variation
Use	No. of people (or visits), by quarter	Same as above	Same as above	Same as above	Same as above	Should provide actual or estimated counts by quarter to allow reporting seasonal variation

Table 2. Recreation areas of the National Park Service

	Type of area	Fed. acres	Non-fed. acres	Total acreage
35	National parks	14,275,000	184,589	14,459,589
13	National historical parks	36,000	8,689	44,689
85	National monuments	9,861,284	355,563	10,216,847
11	National military parks	30,394	1,588	31,982
1	National memorial park	69,528	907,069	976,597
5	National battlefields	2,742	1,486	4,228
4	National battlefield parks	8,022	1,060	9,082
3	National battlefield sites	775	10	785
40	National historic sites	8,333	887	9,220
7	National seashores	231,758	213,246	445,004
3	National lake shores	7,808	67,912	75,720
13	National recreation areas	3,595,102	213,901	3,809,003
21	National memorials	5,487	178	5,665
TOTAL 241		28,132,233	1,956,178	30,088,411

Table 3. Components of recreation availability index

General type of recreational resource	Number	Acreage	Mileage	Visits
National parks	X	X		X
National monuments	X	X		X
National historic sites	X	X		X
National seashores & lakeshores	X	X	X	X
Other national areas	X	X		X
State parks	X	X		X
Urban parks	X	X		
Shorelines & beaches	X		X	X
Wetlands	X	X		X
Wilderness areas	X	X		X
Trails	X		X	

It is clear that spatial availability can also be obtained for a geographical region by summing S_{ij} over all sites j that are in a given region. For example, this would be useful on a broad basis (e.g., West, Northeast) for national parks and for specific states and counties for state and urban parks.

Linear space availability for lakeshores, beaches, and trails would be obtained in a similar fashion, as shown in equations (3) and (4):

$$L_i = \sum_{j=1}^{n_i} L_{ij}, \tag{3}$$

and

$$L = \sum_{i=1}^{n} L_i. \tag{4}$$

The specific i values would refer only to shores and trails.

These aggregations would be useful if reported annually on an absolute basis; they could also be reported as annual percentage changes or referenced with respect to a given year, the initial value being referenced as 100 or any other suitable number. The details for these forms of presentation are not illustrated explicitly.

Visits. Visits to recreational areas would be aggregated for those types of recreation areas indicated in Table 1 and would include:

$$V_{ik} = \sum_{j=1}^{n_i} V_{ijk} \quad \text{(visits to area type } i \text{ in quarter } k\text{)}. \tag{5}$$

By appropriate accumulation, one could also compute:

V_i (annual visits to area type i),

V_k (visits to all areas in quarter k, excluding urban parks),

and

V (annual visits to all areas, excluding urban parks).

Intensity. A measure of intensity during a given quarter for each site j is

$$I_{ijk} = \frac{V_{ijk}}{f_{ij}S_{ij}}, \qquad (0 < f_{ij} \leqslant 1) \tag{6}$$

where f_{ij} is a factor that defines the fraction of the total area of site j that is developed for recreation purposes. This factor is likely to be very nearly 1 for small areas and relatively small for large areas that have only a small portion of their area developed for recreational use. An index of intensity for a particular year is:

$$I_{ij} = \sum_{k=1}^{4} I_{ijk}. \tag{7}$$

For wilderness area, high intensity is not compatible with the characteristics of the area. The interpretation of desirable intensity levels for various types of recreational areas is a value judgment. The average intensity for a type of area i may be obtained by averaging all intensity values for each site.

Aggregate Index of Visits. To obtain an index of visits for a recreation area type i for any subset of types, a weighting factor based upon intensity for each site j will be employed. If

$$I_{ik} = \sum_{j=1}^{n_i} I_{ijk} \, ,$$

then the weighting factor is:

$$W_{ijk} = \frac{I_{ijk}}{I_{ik}} \, . \tag{8}$$

Aggregate indices of visits based upon intensity weighting, analogous to those described earlier, can then be developed.

Recreational Resource Units

A problem that requires further research is the development of a procedure to assign a figure of merit, in the form of a recreation resource unit, to each recreation site. The various types of sites must be categorized into relatively homogeneous groups for this purpose in order to assign appropriate factors to each group and to develop consistent weighting schemes and value scales for specific features of a recreation area. For example, the presence of water in a park used for activities such as camping, picnicking, and outdoor sports enhances significantly the aesthetic desirability and functional use of the facility. This particular feature is less relevant for areas whose primary use is for hiking and has no real significance for historic sites, museums, zoos, and national monuments.

The various factors to be used in rating a given site, not all of which are important to every category of recreation site, include:

1. Diversity of facilities and recreational activities
2. Number of facilities
3. Investment in facilities for various types of activities
4. Size of the site
5. Capacity and utilization
6. Use charges
7. Special features of the area (e.g., presence of water, unique natural features)

Each of these factors must be defined, particularly with reference to the category of site being considered. Problems of value judgments must be considered in weighting the relative importance of these factors as well as in assigning values to the various elements that characterize each factor (e.g., diversity, uniqueness of special features).

If a recreational resource unit could be developed for each site, then an aggregate index of recreational sites could be compiled on a geographic basis by computing an average of units for each site within the geographical region. These averages could be developed for each category individually. An aggregate index for a group of categories could also be computed by applying weighting factors to each category, which might be based upon:

1. Relative number of sites within each category
2. Relative acreage within each category
3. Relative number of visits to all sites within each category

Urban Area Recreation Indices. The indices discussed previously apply to specific recreation sites and to categories of sites. From a sociological point of view, it would be desirable to obtain an index of recreational resources associated with specific urban areas. A number of cities should be selected for this purpose, based on a stratified sample according to population. Figure 1 shows a cumulative distribution of population plotted against cumulative percent of the standard metropolitan statistical areas (SMSA's) for 219 SMSA's ranked in decreasing order of population. The left hand ordinate is based upon the population of the SMSA's being set at 100%. The right hand scale shows the cumulative percent of the U.S. population covered by the representative SMSA's. A feasible method for selecting the N sample locations is to choose the 50% point on the U.S. population scale, which corresponds to about 34% of the

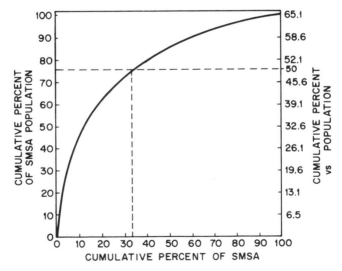

Fig. 1. Cumulative population distribution vs. cumulative percent of SMSA for 219 SMSA's ranked by decreasing population.

SMSA's (approximately 75 of the top 219 SMSA's). Hence, 50% of the sample size N would be selected from these locations. For N = 150, for example, all 75 would be selected. About 15% of the remaining U.S. population is accounted for by the remaining 66% (144) of the SMSA's. For N = 150, one would select about 22 (or about every 7th in decreasing rank order) of the 144 remaining SMSA's. The remaining 53 locations would be chosen from cities not within SMSA's on the basis of combined consideration of size and geographical coverage. This selection could be modified to insure broad geographic coverage.

Recreation Indices for Individual Cities. For any city, m, selected in this manner, a recreation index would be computed, based upon parameters illustrated in Fig. 2 (circular areas around a city, C_m, are depicted for convenience). The distance r_{1m} represents the mean radius from the center of C_m to its boundary. The area of the city itself is depicted as A_{1m}. The distances (from the center of C_m) r_{2m} and r_{3m} (corresponding to incremental areas A_{2m} and A_{3m}) represent a 1-hour drive and a 3-hour drive from the outer boundary of the city. Consideration of the highway system around each urban area would distort the circular contours. The recreation areas are depicted by an x. Indices of recreation for C_m are computed by accumulating the total area of all sites

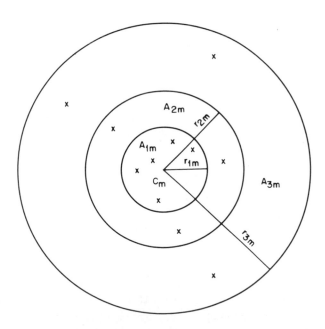

Total Area $A = A_{1m} + A_{2m} + A_{3m}$

Fig. 2. Outdoor recreation areas around city, C_m.

within each area (A_{im}) around C_m, weighting this inversely by the access time from the center of the city, and summing these terms for each incremental area. Let S_{jm} represent the area of site j, which is contained in the total area (A) around C_m. If one assigns a single hour within A_{1m}, then the times corresponding to A_{2m} and A_{3m} are $t_{2m} = 2$ hours and $t_{3m} = 4$ hours, respectively. This implies a normalized set of weights as follows:

$$w_{1m} = \frac{4}{7}$$

$$w_{2m} = \frac{2}{7}$$

$$w_{3m} = \frac{1}{7}$$

If R_{im} is

$$R_{im} = \sum_{j \in A_{im}} S_{jm},$$

then the index of recreational resource for city m is defined by:

$$R_m = \sum_{p=1}^{3} w_{pm} R_{pm} . \tag{9}$$

This represents an index of area of recreational resource for a given city within specific access distances weighted inversely by access distance. Hence, the recreation acreage closest to the core of the city receives the highest weight. Changes in this index would reflect changes in number of recreation sites, size of sites, and access systems associated with each city. Instead of assigning a time of 1 hour to the inner core of a city, one could determine an appropriate time for each city, m, and obtain weights using $t_{2m} = t_1 + 1$ and $t_{3m} = t_1 + 3$.

If a standard area for urban parks is assumed to be 1 acre per 100 people,[6] another form of a recreation index for city m relative to urban parks within A_{1m} may be obtained by defining

$$R'_m = \frac{\sum\limits_{j \in A_{1m}} S_{jm}}{P_m}, \tag{10}$$

where P_m is the population of city m (in 100's of people). A transformation of this index to a 0 to 1 scale could be employed in order to interpret the result as a recreation quality index for parks within urban area boundaries.

Aggregate Recreation Index for a Group of Cities. A population weight for city m in geographic region G is designated by W_m and is equal to the ratio of the population of the city to the population of the region. An aggregate recreation index for region G would be designated as

$$R = \sum_{m \in G} W_m \, R_m \, . \tag{11}$$

This index could be reported in terms of percentage change from year to year or in terms of a base year value set equal to 100.

If a recreation resource unit could be developed for such recreation sites (as discussed earlier), then the areas of the sites would be replaced by the numerical value of the unit, and R_m and R would be developed in terms of a standard recreation unit.

LAND USE INDICES

Shifts in land usage are becoming of more public concern as a result of expanding development and urbanization which leave fewer acres remaining of "unused" land. Although the land resources of the United States are among the richest in the world, there is no national land use policy, and land use data are somewhat difficult to obtain. At present there is very little coordinated federal monitoring related to land use or land quality. Of the 2.2 billion acres of land in the U.S., approximately one-third is publicly owned. The largest single amount, approximately 450 million acres (20% of total land), is administered by the Bureau of Land Management in the Department of the Interior.[13]

Existing Data

Existing widespread land use data can be obtained from the following sources:

1. Agricultural Stabilization and Conservation Service[1] of the USDA has aerial photographs covering 80% of the United States and ranging in age up to 8 years.
2. Conservation Needs Inventory Committee of the USDA[3] has inventoried some two-thirds of the country (non-federal, rural land) in the most recent survey. These data include field observations useful for land categorization.
3. U.S. Geological Survey of the USDI has geologic maps (rock formation, topography, etc.) covering about a third of the country with an additional 1 to 2% coverage per year.
4. Forest Service of the USDA conducts the Timber Resources Survey[17] about every 10 years (most recent, 1965) for the nation's forest lands, which comprise approximately one-third of the total land. Data include photo interpretation and field observations.
5. The Federal Highway Administration of the Department of Transportation publishes annual statistics on approximately 3.7 million miles of highways, roads, and streets at the federal, state, and local level.[10]

The land use data from the sources certainly must overlap, but the degree of overlap is not easily ascertainable. In addition to the preceding agencies, military aerial photos might provide significantly useful information. The data described have wide geographic coverage but are not useful for computational purposes until land use information is codified from aerial photos.

Data that are more immediately useful might be obtained for urban areas, though geographic coverage would not be as widespread. A number of studies have been conducted on land use in specific cities or urban areas. These data might be obtained from the cities themselves, from data submitted to HUD by cities in support of Model Cities applications, or from the Census Bureau. The Neighborhood Environmental Evaluation and Decision System (NEEDS) program of the Bureau of Community Environmental Management in HEW might have data, although this program has not had enough time to accumulate very much information relating to land use. The Geological Survey is undertaking a 27-city project to study the feasibility of using aerial photography to determine land uses and growth patterns by census tract.[18]

Various regional studies might also be utilized. These data could be obtained through reports in the *Journal of the American Institute of Planning* (AIP). Ultimately, it might be necessary to contact local agencies. For example, Santa Clara County (which includes the city of San Jose) in California is known to keep rather complete land use data.[11]

Problem Areas

It has been traditionally the prerogative of the private owner to do with his land as he wishes, and it is the fact of private ownership that makes any type of governmental coordination or control problematical. Nevertheless, it is in the public interest to coordinate land use data so that changes will become obvious and so that planning decisions can be made more intelligently.

A necessary first step in the categorization of land and the measurement of acreage is to provide a suitable land classification scheme. A "Standard Land Use Code Manual" was proposed in 1965 by HUD and the Bureau of Public Roads (then in the Department of Commerce). It was suggested by Clawson et al.[5] that this essentially urban classification scheme be expanded to include rural areas. These rural areas include cropland, forestland, most parkland, wilderness areas, wetlands, and other lands which provide food, timber, mineral, recreational, and wildlife resources. Unfortunately no uniform method of coding land uses has yet been formulated. Moreover, standard category definitions must be developed concurrently with classification and codification.

To illustrate the problems of definition that will arise, certain categories will be discussed briefly:

1. *Urban:* Is a particular municipality greater than a certain size? Population density? Incorporation status?

2. *Building or structure:* Should density of occupancy or floor area ratio be

accounted for? Is a bridge a structure, or is it considered in the paved category? How should buildings be classified which are both residential and commercial?

3. *Paved:* What is minimum width for consideration? What about unpaved driveways? Racetracks? Above ground rights of way? Cemeteries (probably should be green)? Astroturf ®? Surface area of highway overpass counts twice.

4. *Green:* What about dirt yards and spaces with no vegetation? What is minimum area for inclusion? Highway dividers may be landscaped and should be included in this category if of a certain size.

5. *Wetlands:* How are the boundaries of wetlands to be defined? High tide? Particular type of plant?

6. *Bodies of water:* When is a stream a stream? Below some width, is it to be considered land? How big must a pond be in order to be designated as a water area?

Similar problems of definition will arise for other categories. However, the definitions must be standardized before meaningful data can be gathered.

Presently there is no nationally coordinated survey of urban land use practices. Some individual cities have their own inventories, but these are for specific purposes and there is a lack of uniformity in terms of definitions, format, completeness, and time (year of survey).

Land Use Shift Matrix

Any land use index must be based upon a land categorization system. As a first cut, MITRE staff members have chosen 19 categories of land (Table 4). These categories include examples of *natural land types* (desert, wetland) and of *human land uses* (residential, cropland), which may or may not reflect the natural land type. While there is no clear break, the first several categories are natural land types and the last few correspond to human, or man-made, uses. The categories have been chosen to address the following two basic questions:

1. Who is responsible for the land use being shifted? For the three suggested categories of ownership — federal, other governmental (state, county, city, etc.), and private — different legal and/or political action may be necessary. Each of the three ownership categories can be broken down into urban and rural.

2. What types of land should one consider especially important: ecologically vulnerable (wetlands, floodlands), valuable (cropland, habitat, real estate, one of a kind), or controversial? Irreversible land use shifts, such as to paved land or commercial land, might be weighted more heavily than relatively reversible shifts such as woodland to cropland.

The land categories that will be discussed in following pages were chosen to focus attention on shifts in the following types of land:

Table 4. Land use shift matrix

	Wilderness areas	Wildlife refuges	Parkland	Flood plains	Wetlands	Forests and woodlands	Grasslands and rangelands	Desert	Cropland–Regular (I–III)	Cropland–Marginal (IV)	Cropland–Not suitable (V–VII)	Minerals (leased & producing)	Land off-limits to desired use	Residential	Commercial	Industrial	Transportation	Public H.E.W. (incl. instit.)	Waste disposal (j=19)	Other	ROW TOTALS (losses)	NET CHANGE	NEW VALUE
Wilderness areas																							
Wildlife refuges																							
Parkland i = 3			42.9										.3						.1		.4	+5.5	48.4
Undeveloped flood plains																							
Wetlands			1.3																				
Forests and woodlands			2.1																				
Grasslands and rangelands																							
Desert			.3																				
Cropland–Regular (I–III)																							
Cropland–Marginal (IV)																							
Cropland–Not suitable (V–VII)			2.2																				
Minerals (leased & producing)																							
Land off-limits to desired use																							
Residential																							
Commercial																							
Industrial																							
Transportation																							
Public H.E.W. (incl. instit.)																							
Waste disposal																							
Other																							
COLUMN TOTALS (gains)			5.9																				

Notes:
1. Diagonal values represent acres at end of previous year.
2. Values are in millions of acres.
3. Numbers shown are fictitious, for example only.
4. New parkland = 42.9 + 5.9 − .4 = 48.4

1. *Flood plains:* Settlement or development risks extensive flood damage, implies expensive preventive measures, i.e., dams, and increases erosion.

2. *Wetlands:* Should be protected because they are spawning and nursery grounds, highly productive (six times more than cropland), and habitat for ecologically and commercially important species.

3. *Developed lands,* since many acres come from agricultural lands, and *paved lands,* because paving is a measure of urban sprawl, and paving causes erosion, high runoff, unsightliness, and poor wildlife habitat.

4. *Forests:* Important for timber products, recreation, wildlife habitat, and flood control.

5. *Parkland:* Needed for recreation.

6. *Urban green space:* Important for health benefit, aesthetics, and wildlife habitat.

7. *Agricultural land:* Although there doesn't seem to be any concern for scarcity at present or in the foreseeable future, there may come a time of shortage because of increased food demand or losses from civilization encroachment, pesticides, radioactivity, or other pollution.

8. *Land off-limits to desired use:* Due to pollution, hazardous spills, salinization, and/or waterlogging.

It is recommended that annual statistics be kept on some of the 19 categories in Table 4 showing the net amount, the shift to, and the shift from each category. Land shifts in other categories (e.g., forests, deserts) may occur slowly enough so as to require less frequent updating. Overlapping categories will cause problems of non-additivity unless acres belonging in multiple categories are separately identified. Another alternative is to classify land only according to its primary use or type, but the classification might raise more problems than would recording of overlap.

Each element shown is A_{ij}, where A represents acres, i represents the row number, and j the column number. A dummy variable, k, will be used as the category index when referring to acres gained or lost or to total acres. The changes and totals may be computed as follows:

$A_{jk} = \sum_i A_{ijk}$ = acres gained by parkland from other sources during the year. The diagonal value ($i=j$) is omitted.

$A_{ik} = \sum_j A_{ijk}$ = acres lost by parkland to other uses during the year. The diagonal value ($i=j$) is omitted.

$A_k (1970) = A_{ij}$ = acres of parkland at the end of 1970, where $i = j$, diagonal element. Then

$A_k (1971) = A_k (1970) + A_{jk} - A_{ik} = 42.9 + 5.9 - .4$
= 48.4 million acres in fictitious example.

It must be noted that the above example is true only if the acreage figures were non-overlapping. The net change is

$$A_{nk} = A_{jk} - A_{ik} = \text{net (n) change during the year.}$$

Table 5 shows data monitoring characteristics for the various land categories. Acreage acquired or lost should be reported each year by ownership for each state; methods of collection of data include field surveys, aerial photos, and court records.

Land Use Indices

It is possible to compute a number of indices formed by aggregating land shifts from various categories, e.g., indices related to agricultural resources (timber, crops), wildlife habitat change, encroachment, urban green.

Habitat change index would record the numbers of habitat acres lost or gained to wildlife. It would include wilderness areas, wildlife refuges, parkland, undeveloped flood plains, wetlands, forest and woodlands, grasslands and rangelands, desert, and urban green. Cropland could be included. By encroachment is meant the growth of urban and suburban areas and directly related land uses. An *encroachment index* would be limited to "man-made" classes including residential, commercial, industrial, transportation, waste disposal, and public health, education, and welfare. These categories probably encompass those changes felt most directly by the combination of population growth and migration to urban areas. One might later include minerals, cropland, and land off-limits to desired uses, since these are also a result of man's activities; but the encroachment index considered here is limited to the more urban uses. An *urban green index* would be useful for showing that portion of land which is "green" as opposed to being occupied by a structure or covered with paving.

The basic data for the encroachment and urban green indices might be reported in matrix form as shown in Tables 6 and 7. Such tables could be considered for each city, for each type of ownership, and for each state. For the time being, detailed data collection and analysis will have to be limited in scope to a few cities for which meaningful consistent data are available.

The problem of overlapping land uses is expected to occur frequently. If a single portion of land is assigned to more than one category, totaling becomes impossible unless a fractional weighting scheme is used. One could allot such acreage to the primary use category for aggregation purposes, and secondary uses and possible tertiary uses would be simply recorded with the primary use data.

Computation of Indices

The values in the boxes in Table 6 show the acreage (in hundreds of acres) shifted from one category to another; e.g., 200 acres of housing were diverted to highway purposes. Net gains and losses are computed for each category, and together with the diagonal term (last year's total value), a new value is

Table 5. Data monitoring characteristics

Parameter	Potential source[a]	Comments
Wilderness areas	FS, NPS, FWS	FS, NPS, and Bureau of Sport Fisheries and Wildlife all have an interest in wilderness areas. At present there is no federal committee which coordinates these data. A private foundation, the Wilderness Society, Washington, D. C., currently keeps tabs on this activity.
Wildlife refuges	FWS	
Parkland	NPS	State and municipal agencies also have an interest here; however, federal coordination is needed.
Undeveloped flood plains	CE, SCS	A uniform working definition is necessary as to the extent of flood plains. Since many flood plains are already developed and protected, starting in 1971, virgin flood plains should be monitored.
Wetlands	FWS, SCS, CG	A uniform working definition of wetlands is necessary. Of the four classes of wetlands (coastal fresh and salt, inland fresh and salt), CG has responsibility for the two coastal classes.
Forest and woodlands	FS, NPS, SCS	There is some overlap in definition between woodlands (with some grazing) and rangelands (with some trees).
Grasslands and rangelands	FS, SCS	Grassland and rangelands can be split into two parts once a definition is agreed upon. There is some overlap in definition between woodlands (with some grazing) and rangelands (with some trees).
Desert	GS	Relatively little effort has been expended towards monitoring or studying deserts. Their value as a wildlife habitat has been almost completely ignored by the general population. The Bureau of Land Management and the Bureau of Indian Affairs have jurisdiction over much of this land.
Cropland – Regular (I-III)	SCS	The three categories of cropland are used to classify much non-developed land in the U.S.; there will be some overlap with other categories such as grasslands, wetlands, and floodplains. These areas of overlap may cause sufficient confusion in totaling so that each acre of land should be counted only once, by its primary use.

Table 5 (continued)

Parameter	Potential source[a]	Comments
Cropland – Marginal (IV)	SCS	Same as above
Cropland – Not suitable (V-VII)	SCS	Same as above
Minerals (leased & producing)	BM, GS	A definition of mineral areas is needed: surface or subsurface area or both? There is abundant data on quantities of minerals but very little data on acres of land used for mining.
Land off-limits to desired use	Various	Land off-limits to man's desired use includes acres of land so affected by pollution, erosion, flood, fire, or other means that the land has become unsuitable for its previous use. Further categorization is needed to separate man-made from natural agents. This amount of land will be small but significant.
Residential	HUD, COMM	Same as above
Commercial	OBE, COMM	Same as above
Industrial	LABOR, COMM	Same as above

Table 5 (continued)

Parameter	Potential source[a]	Comments
Transportation	DOT	Transportation acres include roads, streets, railroad tracks, bus terminals, airports, stations (not warehouses), and their rights of way. A definition of a road is needed, e.g., must it be paved? Note that transportation acres must logically include parking lots, although they are usually found on private property. Private parking lots will be counted in the "paved" part of the urban green matrix; public lots under transportation.
Public H.E.W. (incl. instit.)	HUD	Public health, education, and welfare areas include hospitals, schools, churches (not parks), museums, and concert halls.
Waste disposal	EPA, BM, DOT	Waste disposal includes landfill, junkyards, and open dumps. A further breakdown to aesthetic and ugly areas is recommended.
Other	Various	The other category is needed to show shifts into and out of the above 19 groups. "Other" does not include all other land.

[a]Restricted here to federal agencies; many state, local and private agencies collect useful data.

FS: Forest Service;
NPS: National Park Service;
FWS: Fish and Wildlife Service;
CE: Corps of Engineers;
SCS: Soil Conservation Service;
CG: Coast Guard;
GS: Geological Survey;

BM: Bureau of Mines;
HUD: Department of Housing and Urban Development;
COMM: Department of Commerce;
OBE: Office of Business Economics;
LABOR: Department of Labor;
DOT: Department of Transportation;
EPA: Environmental Protection Agency.

Table 6. Encroachment index categories, sample chart

	Residential	Commercial	Industrial	Transportation	Public H.E.W.	Waste disposal	Other	Row totals (losses)	Net change	New value (end of year)
Residential	120[a]			2				2	+18	138
Commercial		40						0	+2	42
Industrial			10					0	+5	15
Transportation				100				0	+7	107
Public H.E.W.					15			0	+1	16
Waste disposal						4		0	+1	5
Other	20	2	5	5	1	1	NA[b]	34	−34	NA
Column totals (gains)	20	2	5	7	1	1	0			

[a]Diagonal term represents acreage in that category at the beginning of the year. Numbers are fictitious and represent "hundreds" of acres.
[b]Not applicable.

Table 7. Urban building-paved green matrix

	Building	Paved	Green	Total
Residential				
Commercial				
Industrial				
Transportation				
Public H.E.W.				
Total				

determined. An encroachment index could then be reported in several different forms:

1. An annual compilation of the acres in each category
2. A summation of these various urban acreages
3. An annual percent change from the previous year for each category of the summation
4. A total percent change computed against some baseline year
5. A ratio of the acreage in one category to that of another category or to the total acreage
6. A percentage or fraction related to a comprehensible land mass, e.g., comparing the total national paved area to the size of Rhode Island.

The urban green matrix does not show land use shifts as such; rather it reveals the status of several major land categories with respect to the condition of the surface of the land. The computations involve nothing more than simple arithmetic to obtain such indices as: total acreage which is paved, green, or structure; fraction which is paved, green, or structure; and totals of one year compared to another year (either to the previous or a baseline year).

The acreage-type indices may be given a "quality" designation if appropriate standards for comparison are promulgated. This would involve a national land policy with zoning standards and acreage requirements for the various classes of land. Very little has been done in this respect. Exceptions include the Fish and Wildlife Service, which has set a goal in terms of desired waterfowl wildlife refuge acreage to adequately sustain the present waterfowl population. The Federal Housing Administration of HUD has devised a land use intensity scale[7] to give guidelines for urban land uses such as living, recreational, transportation, and open space. Also the Forest Service has projected how much timber land will be necessary to provide lumber for future housing and other needs.

In computing the indices, it may be discovered that there needs to be more refinement of the categories in order to minimize the effects of overlapping of multiple land uses.

Importance weights can be assigned to the various land use categories and shifts. Irreversible land shifts, such as to paved land or to commercial land, might be weighted more heavily than relatively reversible shifts such as from woodland to cropland. Any weighting scheme would be highly subjective and politically sensitive. Consequently, the first analysis should be restricted to acreage shifts without considering weighting factors. The results of the index computations and the values of the various categories for the selected cities might suggest standards for green space ratios, transportation/total area ratios, etc.

Without any subjective value judgments as to the worth of a given land use, these indices are limited in their interpretability. Time series can show trends of relative acreage, and within limits, the curves may be extrapolated to determine if some threshold level is being approached.

General Recommendations

In order to obtain and maintain land use data on a national scale, it is recommended that:

1. An adequate land use coding scheme be developed,
2. Land category definitions be standardized, and
3. A National Land Use Survey be conducted.

Steps (1) and (2) are interactive. Both must precede step (3). The coding scheme must be expanded for rural areas. Standard definitions either do not exist, or several agencies have their own slightly different versions. A set of uniform interagency working definitions (not necessarily legal, but for index purposes) must be agreed upon. The Land Use Survey might well utilize data and methods from existing programs detailed above. However, a new aerial photography survey of the country might be warranted. Recently, Mexico initiated plans for a total land survey by aerial color photography,[14] to be performed over a period of 12 years.

The Land Use Survey should note ownership (federal, other governmental, or private) and distinguish between urban and rural areas. Ownership data cannot come from aerial photographs, but local property records can be utilized. Of particular interest in urban areas is the amount of acreage that is green, that is paved, and that is building or structure. These data should be obtained in as fine a breakdown as possible.

SUMMARY

This paper has presented concepts for developing indices for outdoor recreational resources and for land use. Indices of recreational availability, visits, intensity, and urban area recreational resources are described. Sources of land use information as well as problems of data assessment also are discussed. The concept of a land use shift matrix is introduced, and several indices which may be derived from such a matrix are proposed.

The material presented in this paper is based upon a study performed by The MITRE Corporation for the Council on Environmental Quality related to a system concept for monitoring the environment of the nation.[8] In addition to analyzing data collection requirements and alternative configurations for processing environmental data, the study described over 100 environmental indices which could be used for: (1) describing environmental status and trends, (2) guiding environmental policy and legislation, and (3) evaluating effects of environmental policies and programs.

REFERENCES

1. Airphoto Use in Resource Management, U.S. Department of Agriculture, Economic Research Service, Information Bulletin No. 336, May 1969.
2. Babcock, L. R. PINDEX, *J. Air Pollution Control Association*, 10: 653-659 (October 1970).
3. Basic Statistics — National Inventory of Soil and Water Conservation Needs, 1967, U.S. Department of Agriculture, Statistical Bulletin No. 46.
4. Brown, R. N., et al. A Water Quality Index — Do We Dare?, presented at the National Symposium on Data and Instrumentation for Water Quality Management, University of Wisconsin, July 1970.
5. Clawson, M. *Land Use Information*, Resources for the Future, Inc., The Johns Hopkins Press, Baltimore, 1965.
6. Clawson, M., and J. L. Knetsch. *Economics of Outdoor Recreation*, Resources for The Future, Inc., The Johns Hopkins Press, Baltimore, 1966.
7. Hanke, B. R. Planning, Developing, and Managing New Urban Areas, in *Soil, Water, and Suburbia*, U.S. Department of Agriculture and U.S. Department of Housing and Urban Development, March 1968.
8. The MITRE Corporation, Monitoring the Environment of the Nation, Report MTR-1660, April 1971.
9. *National Parks and Landmarks* (as of January 1, 1970), U.S. Department of the Interior, National Park Service, 1970.
10. *Newsletter*, Federal Highway Administration, Department of Transportation, Nov. 13, 1971.
11. 1965 Land Use Inventory Manual, Santa Clara County Planning Department, San Jose, California, August 1967.
12. 1970 National Environmental Quality Index, *National Wildlife Magazine*, October—November, 1970.
13. *Public Land Statistics* — 1969, U.S. Department of the Interior, Bureau of Land Management.
14. Quarterly Stockholders' Report, Eastman Kodak Company, February 1971.

15. *Statistical Abstract, 1970,* U.S. Department of Commerce, Bureau of the Census.

16. Thomas, W. A., L. R. Babcock, Jr. and W. D. Shults. Oak Ridge Air Quality Index, Report ORNL-NSF-EP-8, Oak Ridge National Laboratory, 1971.

17. Timber Trends in the United States, U.S. Department of Agriculture, Forest Service Report No. 17, 1965.

18. Wray, J. R. Census Cities Project and Atlas of Urban and Regional Change, presented at the Third Annual Earth Resources Program Review, Houston, Texas, Dec. 1–3, 1970.

BIBLIOGRAPHY

Outdoor Recreation

Outdoor Recreation in the National Forests, Forest Service, USDA Information Bulletin No. 301, Sept. 1965.

N. Dee and J. C. Liebman, A Statistical Study of Attendance in Urban Playgrounds, *Journal of Leisure Research*, 3:2 (Summer 1970).

Data Processing Documentation for Master Inventory File of Existing Public Outdoor Recreation Areas, Bureau of Outdoor Recreation, U.S. Department of the Interior, Dec. 1966.

Recreation – Civil Works Projects, Corps of Engineers, Department of the Army (1967).

R. B. Handley, J. R. Jordan, and W. Patterson, An Environmental Quality Rating System, Northeast Region Staff Paper, Bureau of Outdoor Recreation, 1970.

L. Issacson and B. L. Peterson, Park and Recreational Facilities – Their Consideration as an Environmental Factor Influencing the Location and Design of a Highway, U.S. Department of Transportation (March 1971).

Selected Outdoor Recreation Statistics – 1971, Bureau of Outdoor Recreation, March 1971.

Land Use

R. A. Clark and S. Pollard. Selected References on Land Use Inventory Methods, Council of Planning Librarians, Exchange Bibliography No. 92, July 1969.

A WATER QUALITY INDEX –
CRASHING THE PSYCHOLOGICAL BARRIER

Robert M. Brown

President, National Sanitation Foundation
Ann Arbor, Michigan 48106

Nina I. McClelland

Program Director, National Sanitation Foundation
Ann Arbor, Michigan 48106

Rolf A. Deininger

Associate Professor, School of Public Health
University of Michigan
Ann Arbor, Michigan 48104

Michael F. O'Connor

Research Associate, Public Systems Research Institute
University of Southern California
Los Angeles, California 90007

COMMUNICATING EFFECTIVELY ABOUT THE QUALITY OF WATER

Use of this somewhat bombastic title stems from our desire to draw critical attention to a process of decision making that in practice has been more of an art than a science. At issue is the long standing need for a uniform method for measuring water quality – a "yardstick" with units that are simple, stable, consistent, and reproducible – and an unambiguous method for communicating this information to everyone concerned.

Over the years, and even today, a decision regarding the "quality" of a lake or stream is made as a judgmental determination by the "water expert." This expert examines values obtained in the field or laboratory and, as a result of these values, determines the suitability of a body of water for an intended use. While his decision reflects a process of judgmental weighting and integration of multiple parameter values, the end result does not lend itself to effective communication.

In ancient days, before the unit of measure, the foot, was devised, the plea of the community was for a consistent, uniform, dependable measuring stick so that when one bought a piece of cloth, a beam of wood, or a plot of land, he could know with assurance what he was getting in return for his investment.

We submit that today the plea of the community is for a unit of measure by which water quality can be judged as a commodity.

As requirements for water pollution abatement are being pursued and as stated objectives for water quality attainment are being judged, there is urgent need for those involved in decision making to be knowledgeably aware of the quality status, and changes in such status, of a given surface water. Just who are these decision makers? The general citizen — the taxpayer, persons elected to political office at all levels of government, administrative and regulatory authorities at all levels of government, leaders in business and industry, and those concerned groups who prod the public and private conscience.

All of these decision makers need objective information on quality status upon which to make value judgments. Popularly used descriptive terms like "pristine," "clean," "dirty," "cloudy," "scummy," "stinky," "foul," "noxious," etc. are vague and uninformative at a time when pursuit of water quality objectives calls for clear and precise understanding. "Dollars spent" and "numbers of wastewater treatment plants constructed" are likewise inadequate and improper for reporting benefits, or lack of benefits, from tax monies used for pollution abatement programs. We can note with satisfaction that back in 1889 Lord Kelvin said, "When you can measure what you are speaking about, and express it in numbers, you know something about it; but when you cannot measure it, when you cannot express it in numbers, your knowledge is of meager and unsatisfactory kind."

Thus, determination of water quality involves a process of value judgment. There are several score parameters that may enter into assessment of water quality. The quality determination is made by a knowledgeable person from an array of figures — i.e., parameter values — which he evaluates.

One may ask, then, "Can such individual professional judgment processes be factored and recombined into an empirical expression of water quality?" We believe the answer to this question is an unqualified "yes!" However, considerable reluctance on the part of many professionals to accept the concept of a general use index is clearly recognized. We respond to this attitude with a challenge to the total professional community to help provide for public understanding and critical scrutiny an objective appraisal of stream quality and progress in attaining it.

WQI — CONCEPT AND OBJECTIVES

A mathematical expression generally described as a Water Quality Index (WQI) was developed at the National Sanitation Foundation during 1970. The composite influence of significant physical and chemical parameters is reflected in the index. Application of the index offers a defined, understandable *unit of measure* which responds to change in quality of water. The index method, by virtue of its function to combine the results of change in parameter levels, reflects a net quality value which can be observed and meaningfully interpreted. The real significance of the net changes in quality is not in the magnitude of change, per se, nor in the size of the numerical value, *but in the fact that the WQI does vary* as a result of significant changes in the parameters entering into its determination.

The index, then, provides an instrument or tool that appropriately consolidates, and presents as a single number, the values of multiple parameters selected to enter into the index formulation. With an understanding of the responsiveness of the tool to changes in parameter values and its ability thereby to reflect changes in water quality, the WQI can be viewed as a major resource in *communication*.

Re-emphasizing, the basic objectives of WQI are two-fold:

1. Making available a tool for dependably treating water quality parameter data and presenting it as a single numerical term; and
2. Promoting utilization of a process for effectively communicating water quality conditions to all concerned.

METHODOLOGY FOR DEVELOPMENT OF WQI

In formulating the WQI, a modification of the opinion research technique known as the DELPHI process was used. A panel of 74 persons with expertise in water quality management responded to a series of four questionnaires, described in detail in a previous publication.[1] Parameter selection, quality curves, and the weighted significance of each parameter in the index were determined by the experts. The resulting index formulation is an additive model, expressed as follows:

$$WQI = \sum_{i=1}^{n} w_i q_i$$

where: WQI is a number between 0 and 100; q_i, the quality of the ith parameter, a number between 0 and 100; w_i, the unit weight of the ith parameter, a number between 0 and 1; and n, the number of parameters.

In responding to the questionnaires, the experts were asked to make judgments with respect to *overall* quality. As a result, eleven parameters were proposed for the index:

dissolved oxygen	temperature
fecal coliforms	turbidity
pH	total solids
5-day BOD	toxic elements
nitrate	pesticides
phosphate	

In the WQI equation, however, $n = 9$. Toxic elements and pesticides are classes of substances rather than single parameters and, therefore, cannot be treated as simple additive terms. In these two instances, there will be maximum limits established for each substance within the two classes. When the level of *any* toxic element or pesticide exceeds its maximum limit, the index will automatically go to zero.

In developing quality curves for the WQI, respondents were asked to assign values for the variation in level of water quality produced by different strengths of each of the nine selected parameters. This was accomplished by utilizing a series of graphs. Levels of "water quality" from 0 to 100 were shown on the ordinate; various levels (or strengths) of the particular parameter were arranged on the abscissa. Resulting curves for dissolved oxygen and turbidity are shown in Fig. 1. In each of the graphs, the arithmetic average of all responses is represented by a solid line; 80% confidence limits are shown by dotted lines.

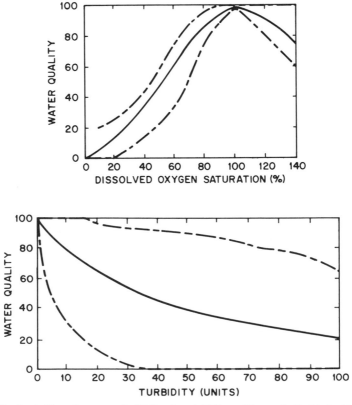

Fig. 1. Arithmetic means (solid lines) and 80% confidence limits (dashed lines) for two water quality parameters.

Convergence was good for some parameters, as shown by the narrow confidence band for dissolved oxygen. However, it varied appreciably for other parameters, as indicated by the wide band for turbidity.

STUDY OF GENERAL VERSUS USE INDICES

In exchanging information with our experts, a number of them continued to be identified with the "use concept hangup." These persons found it extremely difficult, if not impossible, to consider quality of a body of water except in relation to its intended use. Scaling of quality for general understanding is therefore not regarded as relevant by them. Since such single use concepts may not be summarily rejected or ignored, considerable attention has been given to exploring the meaningfulness of single use indices as an alternative to the general purpose index.

A number of disadvantages are perceived with use—related indices: parameters, weights, and scales will vary for each of the large number of uses; much more data will be required to support the additional parameters to be measured; greater expense will be incurred; and communication processes with the public will become commensurately more complex.

Despite these drawbacks — effort, expense, and loss of simplicity — the relative merit of use-related indices versus an index of overall quality was addressed by O'Connor in a doctoral dissertation.[2] He interviewed in person representative members of the panel of experts originally contacted by Brown et al.[1]

One of O'Connor's basic purposes was to probe the methods by which water experts selected parameters to enter into quality determinations, assigned weights (drew curves) for judging variation in the individual parameter values, and rated the parameters relatively in terms of importance. He sought to ascertain whether, in expert decision-making, failure to assign and use parameter weights may result from the complexity of sheer numbers of items to be judged. Might some form of mathematical or statistical appraisal of multiple values be more advantageous than utilizing overall judgments of a large number of dimensions?

O'Connor found that a truly crucial factor lies in identification of the critical parameters which would enter into an index determination. Selection of parameters was ascertained to be of greater importance than determining the shape of the quality curves for designated parameters. Accepting the critical nature of parameter selection, the following factors were identified as having importance:

1. Solid consensus among water evaluation experts as to those parameters that generally and consistently relate to determination of water quality, i.e., those that would be *regularly* required in routine examination of waters in streams and lakes;
2. Parameters for which standard methods are available for routine field sampling and for field or laboratory examination of samples;

3. Parameters that will consistently be influenced by introduction of water pollution factors or by removal of such factors;

4. Parameters that are independently variable, that are not redundant indicators of the same pollution factors, and that do not parallel the indications given by another parameter;

5. Parameters that, from the above four standpoints and the additional considerations of expense and practicable management, will constitute the ingredients for the *least common denominator of water quality* for all water requirements.

Using procedures developed in the study of multidimensional utility analysis, O'Connor developed two water quality indices: (1) for a surface body of raw water to be used to sustain a *fish and wildlife* (FAWL) population and (2) for a water source to be treated and used as a *public water supply* (PWS). The goal of this study was to develop valid indices for two quality-demanding uses of water and to determine the significance of the difference in numbers assigned samples of water with their application. By correlational analysis, the values assigned with FAWL and PWS were compared with those for the overall WQI. Results are shown in Table 1. It is apparent from Table 1 that the WQI correlated better with FAWL and PWS than the two use-related indices did with each other. Thus, O'Connor concluded, "It appears that the WQI is a sort of mean approximation to the PWS and FAWL indices."

Table 1. Correlations among three water quality indices

Index identification	Correlation with FAWL	Correlation with PWS	Sample set no.[a]
WQI	.879	.684	1
FAWL	1.000	.670	1
WQI	.744	.854	2
FAWL	1.000	.667	2
WQI	.792	.652	3
FAWL	1.000	.501	3
WQI	.863	.860	4
FAWL	1.000	.733	4

[a]Sets 1 and 2 are from hypothetical data; sets 3 and 4 are from real data.

It is interesting to note that four parameters are common to each of the indices: dissolved oxygen, pH, nitrates, and turbidity. Five-day biochemical oxygen demand (BOD) and fecal coliforms in the WQI are not included in FAWL; conversely, phenols and ammonia in FAWL are not used in the WQI. However, n = 9 in both WQI and FAWL, and O'Connor concluded that "the WQI can be fairly well described as a FAWL index with fecal coliforms added."

Table 2. Index parameters for three water quality indices

WQI	FAWL	PWS
Dissolved oxygen	Dissolved oxygen	Dissolved oxygen
Nitrate	Nitrate	Nitrate
Turbidity	Turbidity	Turbidity
pH	pH	pH
Temperature	Temperature	Fluoride
Phosphate	Phosphate	Hardness
Fecal coliforms	Ammonia	Fecal coliforms
BOD (5 day)	Phenols	Phenols
Total solids	Dissolved solids	Dissolved solids
		Chloride
		Alkalinity
		Color
		Sulfate

Thirteen parameters are included in the PWS index, the higher number reflecting economic and aesthetic considerations which are not pertinent to WQI or FAWL (Table 2).

Deininger and Maciunas[3] extended the study of Brown et al.,[1] further querying members of the WQI panel, in developing a specific water quality index for public water supplies. Query and analysis of responses followed the processes of the earlier work. Similarities of response were noted with those defined by O'Connor.[2]

Analysis leading to index computation (for public water supplies) was made on data derived from the parameters selected for the WQI (in the report of Brown et al.) and on 11-parameter and 13-parameter groups derived from the further querying of the expert panel. Index formulation was accomplished by use of the arithmetic equation of Brown et al.[1] as well as by use of a specially devised geometric equation. Experience with the geometrically derived index provided additional perspective with regard to the ranging of values on the index scale.

It is interesting to note the conclusions of Deininger and Maciunas[3] after their explorations had been completed:

"1. It is possible to develop a water quality index for surface waters designated for the specific use of public water supply ...

2. The comparisons show that this index developed with a specific use-orientation does not seem to rate water quality levels in a manner markedly different from the rating made by a general, non-specific use-oriented index. Thus, it is possible to argue that water of a certain quality retains that relative quality rating regardless of the use for which it is being considered. Hence, waters of different streams can certainly be compared with regard to changes in quality levels, using one uniformly applied rating scheme.

3. Instead of developing a number of indices for the many water uses, it appears to be more meaningful to further develop and refine a sensitive and general water quality index."

This further study is encouraging and strongly supportive, both with respect to the responsiveness of an appropriately devised index to changes in parameter values and to the sensitivity of a general water quality index in projecting variations in quality irrespective of use orientation.

FIELD EVALUATION

Further evaluation of the effectiveness of the WQI is currently underway in a field sampling program, with the National Sanitation Foundation serving as the coordinating agency. The Michigan Water Resources Commission, Grand River (Michigan) Watershed Council, Maryland Department of Water Resources, Pennsylvania Department of Environmental Resources, California Department of Water Resources, Tennessee Valley Authority, U. S. Army Environmental Hygiene Agency (Edgewood Arsenal), Larimer County (Colorado) Health Department, and the University of Notre Dame Civil Engineering Department are participating in this program. Use problems that may be attributable to

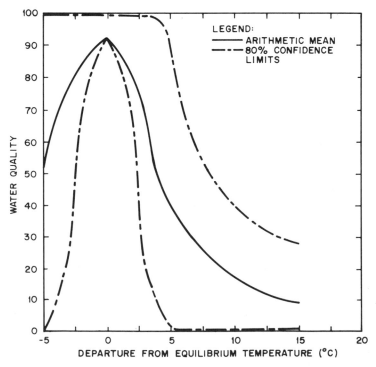

Fig. 2. Arithmetic mean (solid line) and 80% confidence limits (dashed lines) of change in water quality with departure from equilibrium temperature (°C).

sampling, choice of parameters, and application of the additive model are being evaluated.

This study began in May and is expected to proceed for a time sufficiently long for effects of seasonal variation to be apparent (6 to 12 months). The initial goal is to evaluate the WQI with no modification of its current form, i.e, "total coliform density" cannot be substituted for "fecal coliforms," etc. Even before the study began, the subject of temperature as a parameter in the index had to be addressed. In order to develop a meaningful quality curve, the experts were asked to consider the effect that degrees of departure from equilibrium would have on water quality. Thus, the average curve developed for this parameter ranged from −5 to +15°C departure from equilibrium (Fig. 2). This somewhat idealistic approach was quite acceptable for developing the WQI, but the need for defining "equilibrium" was immediate when faced with a field sampling program. By general consensus of the participants, it was agreed that two temperatures would be recorded for every field sampling station − one at the station for measuring degrees of departure and a second, at some point upstream where any effects of heated or cooled discharge were known to be absent, to be called "equilibrium." No problems have been known to result from using this approach.

FURTHER NEEDS

Methods for presenting the finalized WQI to the professional community and the public are currently being reviewed. There is need for innovative educational techniques to help communicate water quality information to the public. Use of a color-coded spectrum for visually depicting use-related applications of the WQI could materially assist in achieving public understanding.

SUMMARY

Determination of water quality is generally regarded as a decision-making process requiring expert judgment. Today this process varies, as different experts apply different weights to various parameters upon which individual determinations of quality are made.

Extended exploration has been made of the need for, and practicability of, a general-use water-quality index that would provide a uniform method for (1) reflecting the quality of water and (2) communicating quality status, and changes in status, to the public. While there is marked reluctance or unwillingness on the part of some water experts to accept a procedure that is not specific use-oriented, significant contra-indications to the feasibility and objectivity of the proposed general WQI have not been encountered during trial uses under field conditions. It would, therefore, seem that obstacles or barriers to the acceptance and use of WQI may be considered psychological rather than technical. We propose to *crash* that barrier.

REFERENCES

1. Brown, R. M., McClelland, N. I., Deininger, R. A., and Tozer, R. G. A Water Quality Index — Do We Dare?, *Water and Sewage Works* (Oct. 1970); also presented at National Symposium on Data and Instrumentation for Water Quality Management, July 1970.
2. O'Connor, M. F. The Application of Multi-Attribute Scaling Procedures to the Development of Indices of Water Quality. Ph.D. dissertation, University of Michigan, 1971.
3. Deininger, R. A., and Maciunas, J. J. A Water Quality Index for Public Water Supplies. Report of a research study, University of Michigan, 1971.

INDICES OF AIR QUALITY*

Lyndon R. Babcock, Jr. and Niren L. Nagda

Associate Professor of Energy and Systems Engineering & Research Associate
College of Engineering
University of Illinois at Chicago Circle
Chicago, Illinois 60680

INTRODUCTION

We have heard much as to the need and utility of environmental indicators and indices. Without them, neither the lay public nor the scientific community can make intelligent decisions with regard to priorities of programs toward environmental improvement.

Our presentation deals specifically with air pollution. First, we will discuss the need for indices and, next, air quality standards from which indices can be derived. Then we will try to show how an index can reorient our priorities. Finally, we will describe a new index proposed as a public information tool and will use it to rank relative air pollution intensities in ten of the nation's cities.

Evaluating overall air pollution can be a complex undertaking. Urban air pollution consists of an often ill-defined mixture of several pollutants emitted from different energy and industrial processes. Additional secondary pollutants are created in the atmosphere. Synergistic interactions can occur between certain pollutants. Despite these complexities, efforts should be made to total the effects of the individual pollutants.

Overall air pollution measures can serve several purposes. The layman wants to know "how bad it is." He may even object to the control agency hiding behind individual part-per-million numbers, which the layman does not understand. Several partial indices are now in use by individual control agencies.

Perhaps more important, a combined air pollution measure or index enables evaluation of the tradeoffs involved in alternative air pollution control strategies or in evaluation of control equipment which, for instance, reduces levels of certain pollutants while increasing levels of others. Programs aimed at emission control of automobile and jet aircraft engines can result in this situation.[7] We must carefully weigh the pros and cons of any proposed control equipment before requiring wholesale installation. Only with a usable total air pollution

*ORAQI was developed within the Environmental Program at Oak Ridge National Laboratory, operated by Union Carbide Corp. for the Atomic Energy Commission, with financial support by the National Science Foundation RANN Program. Additional support was supplied at the University of Illinois at Chicago Circle by the National Science Foundation (Research Grant No. GK-27772) and by the U.S. Department of Transportation.

Table 1. USA air pollution source distribution (1969)

Source	PM	SO$_x$	NO$_x$	CO	HC	Total
Uncorrected basis (10^6 ton/year)[a]						
Transportation	0.8	1.1	11.2	111.5	19.8	144.4
Fuel combustion (stationary sources)	7.2	24.4	10.0	1.8	0.9	44.3
Industrial processes	14.4	7.5	0.2	12.0	5.5	39.6
Solid waste disposal	1.4	0.2	0.4	7.9	2.0	11.9
Miscellaneous (Forest and agricultural fires, etc.)	11.4	0.2	2.0	18.2	9.2	39.2
Total	35.2	33.4	23.8	151.4	37.4	281.2
Weighted basis (% of grand total)						
Transportation	0.9	1.0	6.9	3.0	4.6	16.4
Fuel combustion (stationary sources)	10.2	19.8	3.7	0.1	1.1	34.9
Industrial processes	20.4	6.1	0.1	0.3	0.1	27.0
Solid waste disposal	2.0	0.1	0.3	0.2	0.2	2.8
Miscellaneous (Forest and agricultural fires, etc.)	16.1	0.2	1.3	0.5	0.8	18.9
Total	49.6	27.2	12.3	4.1	6.8	100.0

[a]Uncorrected data from Ref. 10.

yardstick can we get the most from pollution control expenditures. In addition, a total air pollution measure enables one to compare overall air pollution from different sources or in different localities. Indices also permit establishment of meaningful long term trends even when the pollutant mix is changing.

The simplest measure of total air pollution unfortunately is in wide use and involves the simple summation of individual pollutant emission weights. A breakdown of a familiar summary of USA emissions[10] is shown in the upper half of Table 1. Quite clearly, when using an uncorrected weight basis, the USA air pollution problem consists largely of carbon monoxide emanating from automobile engines. Note that oxidant does not appear on Table 1. Worse, relative toxicities are not considered in gross weight comparisons.

AIR QUALITY STANDARDS

The top half of Table 1 is a totally unsatisfactory assessment of the national air pollution problem simply because some air pollutants are more important than others. That is, at a given concentration, some pollutants are more toxic or more unpleasant. To further complicate the problem, pollutants have different effects. For instance, particulates affect visibility, a safety and esthetic effect; sulfur oxides have an unpleasant odor, are corrosive to metals and other materials, and adversely affect the human respiratory system; oxidant can irritate eyes and kill certain kinds of plants. Standards then are needed to address at least three different adverse effects:

> health
> esthetic
> economic

Another difficulty is the varied, uneven response of individuals to various pollutants. From a health standpoint, the very young and the very old are most sensitive to air pollution dosage. Then there is the seemingly irrational response. For instance, the wealthy chain-smoking hilltop home owner would seem more concerned about his view of a distant mountain or the city lights at night than he is about his own health and life expectancy.

At this point most practitioners, recognizing these and other difficulties, opt toward control programs aimed at reducing or maintaining each of the offending pollutants below some specified level. These levels are called ambient air quality standards, levels above which adverse effects (of some kind) are known or thought to occur. These ambient standards must be differentiated from emission standards. The latter are used to govern pollutant sources such that ambient concentration levels will be maintained below their adverse values.

How then are ambient standards set? In the past, several states and localities have set standards based on sketchy medical and crop damage evidence, and often the final level adopted was based on a negotiated compromise involving both health considerations and the cost of available control technology.

The U.S. Environmental Protection Agency (EPA) has recently established standards for six pollutants.[15] The levels are largely based upon criteria documents published earlier.[1-5]

The purpose of this paper is not to evaluate these standards. Rather, despite their admitted deficiences, they are taken as given, as the best available approximations for the adverse threshold levels of the individual pollutants. These standards, listed in Table 2, were extrapolated to a common 24-hour-averaging time basis using the method proposed by Larsen[12] and later

Table 2. EPA ambient air quality standards adopted April 30, 1971[a]

Pollutant	Annual mean	Levels not to be exceeded more than once/year			
		24 hr	8 hr	3 hr	1 hr
Oxidant, ppm					0.08
Sulfur oxides, ppm	0.03	0.14			
(secondary)	0.02	0.10		0.5	
Nitrogen dioxide, ppm	0.05				
Carbon monoxide, ppm			9		35
Particulate matter, $\mu g/m^3$	75	260			
(secondary)	60	150			
Hydrocarbons, ppm				0.24[b]	

[a] Data from Ref. 15
[b] 6 to 9 AM only

Table 3. Comparison of extrapolated standards with background
and urban pollution concentrations

Pollutant	Standards (24 hr)		Concentrations[a]	
	$\mu g/m^3$	ppm	background[b]	urban[c]
Oxidant (OX)	59	0.03	0.02	0.03
Particulate matter (PM)	150	-	37	120
Sulfur oxides (SO_x)	266	0.1	0.0002	0.05
Nitrogen dioxide (NO_2)	400	0.2	0.001	0.04
Carbon monoxide (CO)	7800	7.0	0.1	7

[a] ppm except $\mu g/m^3$ for particulate matter

[b] Data from Ref. 9

[c] 1966 annual averages for CAMP cities: Chicago, Cincinnati, Denver, Philadelphia, Saint Louis, Washington (Ref. 6).

partially confirmed by McGuire and Noll.[14] Standards thus derived were used as the basis for the indices discussed herein. Where several standards exist for a single pollutant, the most stringent was selected. These extrapolated selected standards are listed in Table 3 (columns 1 and 2).

Standards proposed for California (1969) were incorporated into an earlier index.[7] Relative to particulate matter, the EPA standards discussed herein have been made more stringent for carbon monoxide (2 times) and oxidant (2 times) and less stringent for nitrogen dioxide ($\frac{1}{4}$ times). The importance of hydrocarbons is also reduced because hydrocarbons are now considered a pollutant precursor rather than a pollutant. According to EPA,[15] "The sole purpose of prescribing a hydrocarbon standard is to control photochemical oxidants."

Note that differences between pollutants are indeed significant (Table 3). At the extreme, a given concentration of oxidant is shown to be 160 times as unpleasant as the same concentration of carbon monoxide.

Table 3 also compares the standards both with the estimates of unpolluted background levels[9] (column 3) and with average concentrations found in urban areas[6] (column 4). The background levels are fairly insignificant for three of the pollutants. However, the reported background level for oxidant almost exceeds the 24-hour standard. Also disturbing are the high, relative to the standards, average pollutant levels in the nation's urban centers. The picture is worse than depicted by Table 3, since annual averages are compared therein with 24-hour standards. Standards extrapolated to a 1-year averaging time are lower than those for shorter averaging times.

Unfortunately, the EPA standards say nothing about combinations of pollutants. Even aside from the synergism question, could 0.09 ppm sulfur oxides and 0.19 ppm nitrogen dioxide when present together (both slightly below their 24-hour standards) be more adverse than 151 $\mu g/m^3$ of particulate matter alone (slightly above its 24-hour standard)? In the indices described herein, such is assumed to be the case: there is no threshold; pollutant effects are additive even when below their standards. The highly approximate nature of the present independent pollutant standards foretells a considerable time period before effects of pollutant combinations can be quantified.

COMBINED POLLUTION INDICES

There is increasing recognition of the need for integrated appraisal of the air environment, beyond levels of single pollutants. Official and unofficial indices are appearing in various localities. Some are discussed in an earlier paper.[7] More recently, Fulton County (Atlanta), Georgia, has instituted an index that reports the levels of three pollutants as fractions of their standards and then sums the contributions to yield a single index number.[11] Unfortunately, Atlanta and most cities do not measure all five of the most important pollutants; most indices tend to emphasize local situations. The proliferation of non-comparable indices is beginning to result in considerable confusion.

Episode plans also can utilize indices. Episodes occur during adverse meteorology when pollutants levels can climb, far beyond the EPA standards, to dangerous concentrations where serious action (usually stagewise shut down of industry and transportation or even evacuation) is required to minimize loss of human life. The federal guidelines for such incidents include use of the product of coefficient of haze and sulfur dioxide as a measure of pollution intensity.[17]

A comprehensive combined air pollution index could serve both needs; it could measure day-to-day changes in total air quality as well as indicate the presence of episodes.

"Pindex" is such an index that was proposed prior to establishment of the federal standards. Although applicable to ambient data, Pindex was created specifically for the study of emissions. Pindex has a sulfur oxides—particulate matter synergism term and a photochemistry provision wherein oxidants are synthesized from nitrogen oxides, hydrocarbons, and solar radiation. The method is described in full and applied to several examples in an earlier paper.[7] This model has now been revised to reflect the 1971 EPA standards: the sulfur oxides—particulate synergism term and the tolerance factor for excess hydrocarbons have been removed; the other tolerance factors were revised to the values shown in column 1 of Table 3. Finally, excess nitrogen oxides have been assumed to be 50% NO_2 for application of the nitrogen dioxide standard. This revised model was then used to adjust the gross emission data shown in the top half of Table 1. The results, reported on the bottom half of Table 1, were normalized to 100% with the oxidant contribution prorated between the nitrogen oxide and hydrocarbon precursors. These results essentially duplicate those reported for the original Pindex model and using an earlier emission inventory:

"Use of tolerance factors plus inclusion of oxidant and synergism terms has completely reordered the USA air pollution problem. Carbon monoxide which dominated the source distribution based on emission weights became almost insignificant after pindexing. Particulate matter became the most serious USA air pollution problem, with sulfur oxides second, and nitrogen oxides a strong third. Despite its essential contribution to photochemical oxidant synthesis, hydrocarbons have assumed a low value only slightly ahead of carbon monoxide."[7]

Among the sources, transportation remains a significant pollution category, but Pindex (revised) lowered transportation to fourth rank, behind stationary combustion, industry, and miscellaneous open burning.

The 90,000,000 automobiles running about the country insult our environment and the public welfare in many ways. Air pollution is just one example; noise, congestion, traffic deaths, and high land use are others. Our government has rightfully chosen to mount a major federal-level attack on these problems.

Scientists have an obligation not only to identify sources of environmental degradation but also, and much more important, to put them into realistic and useful perspective. The application of a combined air pollution index tends to lessen the importance of the automobile relative to other sources of air pollution.

Pindex (revised) doesn't tell the whole story, of course. Lead, odor, and specific toxic or even carcinogenic hydrocarbons were not addressed. Nor was the distribution of sources and the proximity of the population considered. With

Table 4. Emissions: transportation sub-category (1969)

Source	PM	SO_x	NO_x	CO	HC	Total
Uncorrected basis (10^6 ton/year)[a]						
Motor vehicles (gasoline)	0.3	0.2	7.6	96.8	16.9	121.8
Motor vehicles (diesel)	0.1	0.1	1.1	1.0	0.2	2.5
Aircraft	0.1	0.1	0.4	2.9	0.4	3.9
Railroad	0.1	0.1	0.1	0.1	0.1	0.6
Vessels	0.1	0.3	0.2	1.7	0.3	2.6
Motor vehicles (non-highway)	0.1	0.2	1.8	9.0	1.9	13.0
Total	0.8	1.1	11.2	111.5	19.8	144.4
Weighted basis (% of grand total)						
Motor vehicles (gasoline)	0.4	0.2	4.8	2.6	3.2	11.2
Motor vehicles (diesel)	0.1	0.1	0.5	0.0	0.3	1.0
Aircraft	0.1	0.1	0.2	0.1	0.2	0.7
Railroad	0.1	0.2	0.1	0.0	0.0	0.4
Vessels	0.1	0.2	0.1	0.1	0.1	0.6
Motor vehicles (non-highway)	0.1	0.2	1.2	0.2	0.8	2.5
Total	0.9	1.0	6.9	3.0	4.6	16.4

[a]Uncorrected data from Ref. 16.

these cautions in mind, let's take a closer look within the transportation category. Table 4 shows the EPA intra-category distribution for transportation,[16] both before and after Pindex (revised) correction. As expected, the "motor vehicles (gasoline)" category dominates transportation emissions, accounting for 68% of the transportation total, but only 11.2% of the grand total. With the surprising exception of non-highway motor vehicles, the other categories make relatively minor contributions. The rapidly growing, largely unregulated, non-highway category presumably includes snowmobiles, trail bikes, outboard motors, etc. and already accounts for 15% of the transportation pollution and 2.5% of the nationwide grand total.

We must also reorient our thinking with regard to the relative importance of specific pollutants emanating from transportation equipment. As expected, particulates and sulfur oxides make a real but relatively small contribution to the transportation category. Carbon monoxide, because of its large volume and stringent standard, is more important. Hydrocarbons, despite lack of a standard in the model, make a large contribution through oxidant synthesis. Nitrogen oxides, however, contribute to oxidant syntheses; at the same time, excess nitrogen dioxide is a serious air pollutant in its own right. Inclusion of this pollutant in control programs seems well justified.

Combined air pollution indices can indicate where our control priorities should lie, both within and outside the transportation category.

OAK RIDGE AIR QUALITY INDEX (ORAQI)

Air quality indices can serve several purposes, and indices constitute one of the areas emphasized within the Oak Ridge National Laboratory–National Science Foundation (ORNL-NSF) Environmental Program. Several efforts are underway toward development of air quality indices. The remainder of my presentation will describe a first product.

The goal was to create a public information vehicle – an understandable, meaningful, and easy-to-compute air quality index suitable for communication between the media and the lay public. Within limits, we decided to sacrifice accuracy and sophistication in order to achieve our primary goals. The result[19] of our effort is the Oak Ridge Air Quality Index (ORAQI) published in September 1971 and widely distributed to TV and radio stations by the National Association of Broadcasters. The Knox County (Tennessee) Air Pollution Control Department began routine use of ORAQI in October 1971, and it is now being used by an increasing number of cities. Eventually, wide use of ORAQI may enable meaningful air quality comparisons between cities across the country.

ORAQI weights and sums the contributions of five of the six pollutants for which standards have been promulgated by EPA:

Carbon monoxide   Oxidants
Sulfur oxides   Particulate matter
Nitrogen dioxide

The sixth pollutant category, hydrocarbons, is not directly included in ORAQI because hydrocarbons are assumed to be a pollutant precursor rather than an actual pollutant. The hydrocarbon contribution is included in the oxidant measurement. The ORAQI system can be easily revised to include additional or revised standards as they are established.

ORAQI is represented mathematically by the following equation:

$$ORAQI = \left(5.7 \sum_{i=1}^{i=5} (C_i/S_i)\right)^{1.37}$$

where

C_i = concentration of pollutant i

S_i = EPA standard for pollutant i

The coefficient and exponent are used to scale ORAQI such that non-polluted background levels (Table 3, column 3) give an ORAQI of 10 and pollutant levels at their limits (Table 3, column 2) yield a value of 100.

Although the exponent is primarily a scaling factor, there is some additional precedent and justification for using one. Larsen[13] derived exponents of 1.5 for sulfur dioxide and 1.9 for particulates when correlating pollutant levels with deaths during episodes. Further, the Ontario index[18] is a weighted sum of coefficient of haze (COH) and sulfur dioxide levels raised to the 1.35 power. Admittedly here, too, the 1.35 is basically a scaling factor that yields an "episode threshold level" when the index reaches 100. Finally, there is the intuitive feeling that an increment of pollution added to clean air is less harmful than the same increment added to already-polluted air.

The heart of the ORAQI system is the easy-to-use nomograph shown in Fig. 1. The first five columns relate the measured pollutant concentrations to their EPA standards (from Table 3, column 2).

To use the nomograph, pollutant concentrations are entered on the right-hand scales of the summing columns. Concentrations corresponding to EPA standards are each equivalent to 10 units on each appropriate left-hand scale. These summing columns thus translate each pollutant concentration into a fraction of the appropriate standard. The translation is analogous to that used in Pindex and in the Fulton County index.

Summation of the left hand units on the five columns gives the measured total to be entered on column 6. If concentrations for all five of the pollutants are known, ORAQI is found at the intersection on column 7 with a line connecting the reading on column 6 with "none" on column 8.

Column 8 permits the inclusion of approximate contributions of pollutants which are not monitored. All combinations equivalent to less than tripling the column-6 level are included on column 8. After evaluating several alternatives, it was decided that the combinations of unknown pollutants be located on column

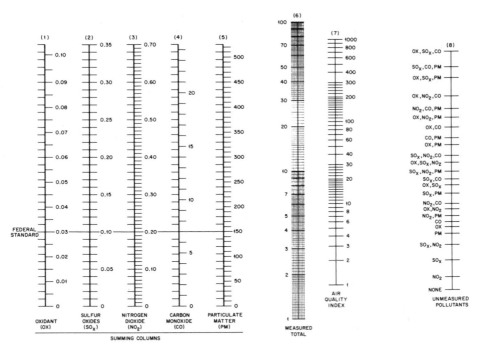

Fig. 1. ORAQI nomograph. Use $\mu g/m^3$ for particulate matter, ppm for other pollutants.

8 so as to provide a fractional increase proportional to pollution levels found in fully-monitored USA urban areas. The percentages used are listed below and were determined by dividing each concentration (Table 3, column 4) by its standard and normalizing the results to 100%. Thus:

USA urban pollution distribution

Oxidant	26.8%
Sulfur oxides	13.8%
Nitrogen dioxide	6.3%
Carbon monoxide	28.7%
Particulate matter	24.4%
	100.0%

For example, if all the pollutants except nitrogen dioxide were monitored, column 8 would add roughly 6% to the measured total. Discretion must be employed; the approximation is usually satisfactory except in instances where the major pollutants are not monitored. For instance, quite erroneous results would occur for Los Angeles should only particulate matter and sulfur oxides be monitored. Column 8 would not satisfactorily estimate the oxidant, nitrogen dioxide, and carbon monoxide contributions.

The application of ORAQI can be further broadened by resorting to another compromise. In many cities, tape samplers (COH or soiling index) are used to assess particulate level. In such instances, 100 times the COH reading can be substituted to approximate particulate loading in $\mu g/m^3$.

The reader may rightfully question the liberties taken herein. We do not argue that levels of certain pollutants can always predict the levels of others (If this were true, a single instrument would suffice for air quality monitoring.) Nor do we defend, in the face of so much conflicting evidence, any simple relationship between the tape sampler and high-volume-sampler results. Where utilized, these expediencies negate city to city comparisons. But because of the

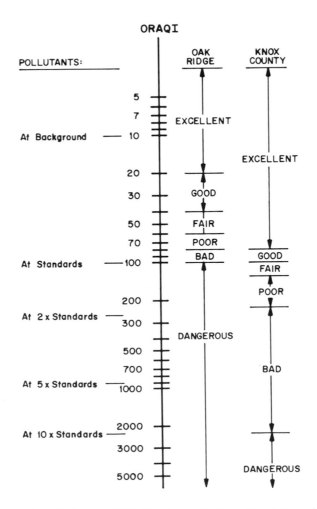

Fig. 2. Relationships between ORAQI scale, standards, and local air quality scales.

dearth of comprehensive monitoring efforts, these compromises do seem justified in order to enable wider use of ORAQI. Further, the availability of, and attempts to use, a meaningful index now may very well promote increased monitoring activity throughout the nation.

This paper stops short of interpreting the meaning of ORAQI levels. To date, we are unable to realistically relate standards to effects upon the "average man." Before labeling ORAQI levels, one must decide whether the 100 level marks the ending of good air and the beginning of deterioration or indicates complete deterioration of air quality. Oak Ridge National Laboratory proposed the latter; Knox County adopted the former. The two systems are compared on Fig. 2.

ESTIMATES OF URBAN AIR QUALITY USING ORAQI

Before closing, it seems appropriate to apply ORAQI to some of the nation's data; if sufficient recent data can be obtained, a future paper will apply ORAQI in an attempt to discover and interpret recent air quality trends within USA cities. In the meantime, the top half of Table 5 lists some data taken in USA cities during the 1962–1967 period.[7] The data should not be construed to

Table 5. Air quality data for USA cities, annual averages[a]

City	PM ($\mu g/m^3$)	SO_x (ppm)	NO_x (ppm)	CO (ppm)	Oxidant (ppm)
Chicago	124	0.14	0.14	12.0	0.01
Cincinnati	154	0.03	0.06	6.0	0.02
Denver	126	0.01	0.07	7.9	0.03
Los Angeles	119	0.02	0.13	11.0	0.05
Philadelphia	154	0.08	0.08	6.8	0.01
Saint Louis	143	0.04	0.07	5.8	0.04
San Diego	69	0.01	0.05	3.0	0.03
San Francisco	68	0.01	0.14	3.2	0.02
San Jose	92	0.01	0.12	5.0	0.02
Washington	77	0.05	0.07	6.0	0.02

ORAQI Levels

City						Total
Chicago	16	27	7	33	7	90
Los Angeles	15	4	6	31	32	88
Saint Louis	17	8	3	14	21	63
Philadelphia	17	14	3	17	6	57
Denver	14	2	3	19	17	55
Cincinnati	17	5	3	14	11	50
San Jose	10	2	5	11	16	44
Washington	8	8	3	14	10	43
San Francisco	7	1	5	11	10	34
San Diego	7	1	2	6	15	31

[a]NO_x assumed to be 50% NO_2; basic data from Ref. 7.

represent present integrated average air quality in these cities, but taken as given, the data can serve to illustrate the application of ORAQI.

Large differences between cities are apparent despite the use of annual averages. Chicago leads in sulfur oxides and in carbon monoxide. Los Angeles has a different problem with low sulfur oxides but has the highest oxidant level. Cincinnati and Philadelphia are highest in particulates. Which city has the highest level of overall air pollution? The question can be answered through use of a combined index.

The bottom half of Table 5 lists the calculated ORAQI levels according to rank and also shows the contributions of each pollutant to the total.

Chicago ranks highest with high levels of all pollutants except oxidant. Air pollution in Los Angeles is nearly as great with the low sulfur oxide level more than compensated by high oxidant.

Despite high levels of particulates, Philadelphia and Cincinnati have intermediate rankings. San Diego benefits from relatively low levels of all the pollutants and falls at the bottom of the list. Note the large gap separating Chicago and Los Angeles from the other cities.

The bottom half of Table 5 also indicates where a city's control priorities should lie. Although the automobile was shown to be a relatively low priority problem based on nationwide emissions (Table 1), the automobile re-emerges as a significant pollution source in urban areas. The carbon monoxide–oxidant combination, largely attributable to the automobile, accounts for over half the total in seven of the ten cities studied. Certain eastern cities still couple an automobile problem with high contributions from particulates and sulfur oxides. In fairness, it should be noted that cities such as Chicago, through fuel regulations, have made considerable progress toward reduction of sulfur oxide levels since 1966. Such reductions are not reflected on Table 5.

The rankings using ORAQI almost duplicate those reported earlier[7] when Pindex was applied to the same data. ORAQI did drop San Jose from fifth worst to seventh and San Francisco from eighth to ninth.

DISCUSSION

Implications

Combined pollution indices enable more objective determination of the contributions of individual pollutants and sources to overall air quality. For instance, Table 1 indicates almost 50% of the nation's air pollution problem still results from particulate matter, despite the long availability of well-tested control technology.

The automobile question is less clear. Automobiles were shown to be a relatively minor source of the nation's overall air pollution. However, the ORAQI comparison of cities indicated that automobiles may be the major pollution source within many cities. Where such questions arise, emissions should be more carefully related to ambient air quality through use of dispersion models.

Cautions

Thus, caution must be exercised when applying indices. The assumptions made and inherent limitations must be recognized. Errors and inaccurate assumptions can be magnified through use of indices.

The indices and examples described herein are based on original data and EPA standards; both are suspect. They were used because they were available; indeed national strategies are being based on this information.

The standards themselves are subject to change, but of course tightening or loosening the standards in concert does not affect a combined index. However, relative changes do affect index results. Such changes in single tolerance factors or standards were subjected to a sensitivity analysis in an earlier paper,[8] and it was found that index results were surprisingly insensitive to changes in the standards. Tables 1 wholly and Table 5 partially support this conclusion. However, Table 5 also indicates that use of the new EPA standards increased the relative importance of the automobile as a pollution source within urban areas. This effect is directly attributable to the very stringent carbon monoxide and oxidant standards established by EPA.

More-sophisticated indices

One often must compromise between sophistication and simplicity. Both Pindex and ORAQI compromise toward the latter.

There are of course many other useful forms of indices which might be pursued. Perhaps dosage is a more important criterion than level. In a related vein, a threshold index could be used as a go—no go test wherein only days that exceed the standard would be counted. More sophisticated are probablistic indices that would relate levels both to standards and to the probability of an occurrence. The succeeding paper discusses one such index, also developed at Oak Ridge National Laboratory. Work in a new area must begin at the beginning, and even ORAQI often seems beyond our monitoring and relating-to-effects capabilities. As we become more sophisticated in these areas, more sophisticated indices can be better utilized.

SUMMARY AND CONCLUSIONS

Combined air pollution indices seem to be an overdue necessity. Formulators of control policy and the lay public must have improved methods for integration of emissions and ambient pollutant levels into more meaningful assessments of air quality.

ORAQI and Pindex have been proposed, and examples using these indices have been discussed. These indices are based upon air quality standards, and the importance, as well as the uncertainty of these standards, has been emphasized.

Despite obvious limitations, combined air pollution indices, which combine generally accepted tenets of air pollution technology, can be useful tools. Such tools become more useful when used in combination with others such as local emission surveys and meteorological dispersion models. The ultimate merit of indices such as ORAQI will become apparent only when indices are put to wider use.

ORAQI is a first step, a general framework that will be updated when its shortcomings become apparent and as new information becomes available. Existing factors and relationships can be revised, and new pollutants can be added as additional EPA standards are established.

Even if the results are questioned, ORAQI may still serve a useful function by helping to identify faulty standards and inadequate monitoring programs.

REFERENCES

1. *Air Quality Criteria for Carbon Monoxide*, National Air Pollution Control Administration, Report No. AP-62. Durham, North Carolina, March 1970.
2. *Air Quality Criteria for Hydrocarbons*, National Air Pollution Control Administration, Report No. AP-64. Durham, North Carolina, March 1970.
3. *Air Quality Criteria for Particulate Matter*, National Air Pollution Control Administration, Report No. AP-49. Durham, North Carolina, January 1969.
4. *Air Quality Criteria for Photochemical Oxidants*, National Air Pollution Control Administration, Report No. AP-63. Durham, North Carolina, March 1970.
5. *Air Quality Criteria for Sulfur Oxides,* National Air Pollution Control Administration, Report No. AP-50. Durham, North Carolina, January 1969.
6. *Air Quality Data, 1966.* National Air Pollution Control Administration, Durham, North Carolina, 1968.
7. Babcock, L. R. A Combined Pollution Index for Measurement of Total Air Pollution, *J. Air Poll. Control Assoc.*, 20: 653–659 (October 1970).
8. Babcock, L. R. A Combined Pollution Index: Effects of Changing Air Quality Standards, pp. 65–81 in *Determination of Air Quality,* G. Mamantov and W. D. Shults (eds.). Plenum Press, New York, 1972.
9. *Cleaning our Environment, the Chemical Basis for Action.* American Chemical Society, Washington, 1969.
10. *Environmental Quality (Second annual report).* Council on Environmental Quality. Washington, August 1971.
11. *Fulton County Air Pollution Index* (mimeo, received 1971, undated). Fulton County Health Department, Atlanta, Georgia.
12. Larsen, R. I. A New Model of Air Pollutant Concentration Averaging Time and Frequency, *J. Air Poll. Control Assoc.*, 19: 24–30 (January 1969).
13. Larsen, R. I. Relating Air Pollutant Effects to Concentration and Control, *J. Air Poll. Control Assoc.*, 20: 214–225 (April 1970).
14. McGuire, T., and K. E. Noll. Relationship between Concentrations of Atmospheric Pollutants and Averaging Time, *Atmos. Environment* 5: 291–298 (1971).
15. National Primary and Secondary Ambient Air Quality Standards, *Federal Register*, 36: 8186–8201 (April 30, 1971).

16. *Nationwide Inventory of Air Pollutant Emissions: 1969*. Environmental Protection Agency, Washington, in press.

17. Requirements for Preparation, Adoption, and Submittal of Implementation Plans, *Federal Register* 36: 15486–15506 (Aug. 14, 1971).

18. Shenfeld, L. Note on Ontario's Air Pollution Index and Alert System, *J. Air Poll. Control Assoc.* 20: 612 (July 1970).

19. Thomas, W. A., L. R. Babcock, and W. D. Shults. *Oak Ridge Air Quality Index* Report (ORNL-NSF-EP-8), Oak Ridge National Laboratory, Tennessee, September 1971.

STATISTICALLY BASED AIR-QUALITY INDICES*

W. D. Shults

Leader, Analytical Methodology Group
Analytical Chemistry Division
Oak Ridge National Laboratory†
Oak Ridge, Tennessee 37830

John J. Beauchamp

Statistics Department
Mathematics Division
Oak Ridge National Laboratory†
Oak Ridge, Tennessee 37830

INTRODUCTION

Oak Ridge National Laboratory is, under sponsorship of the NSF, studying the development of air quality or air pollution indices. We have examined a number of ways for combining measured information about several pollutants into a single number that reflects the overall state of the air quality within a given area. A number of these index systems have been strictly mathematical relationships that tended to give the index itself desirable qualities, that is, the desired response as values of the measurements changed. As this work progressed, however, we have come to believe that some sort of statistically based index is probably the ultimate answer. We feel this way for several reasons, one of which is that there seems to be some truth in the belief that air quality measurements follow a definable statistical distribution. More importantly, the hazard or harm that results from air pollution is usually measured – when it is measured – in statistical terms, that is, the frequency of effec   it seems that if we could devise a valid statistically based indexi   should be compatible with the damage information or dat   hat will become available in the future. Accordingly, s   thing to say about   standards are set or at   nt of standard

rationale a

Indeed w

complete

and ref

---

*Re

†O

thoughts, suggestions, ideas, and criticisms of others as feedback on our development process.

ASSUMPTIONS

We have had to make a number of assumptions in this work and we want to mention them first. First of all, the assumption is made that the necessary data are available. This, of course, is not always true. We don't always have the necessary data, but we could. A valid index should have something to say about the types and amounts of data that are needed.

Assumption number two is that standards have been set and are valid. This is tantamount to saying that we accept the existing standards whatever they may be and whether we like them or not. They are the best that we have. They may be subjective to a certain extent, but they are agreed upon. A proven index, however, may reflect on the inadequacy of these standards – may indicate where change or reevaluation is needed and ultimately may indicate how better standards might be set.

Assumption number three is that a suitable statistical distribution can be found for air quality data. In other words, the distribution of air quality measurements can be described by certain parameters such as a mean and a standard deviation, with a statistically acceptable distribution. In Tables 1 and 2, we present means and standard deviations for the distributions of four major air pollutants in five major cities, for the two cases when the air quality

Table 1. Mean (standard deviation) – 1968 CAMP data. Normal assumption.

	Denver	Philadelphia	Cincinnati	Chicago	St. Louis
Carbon monoxide	5.34(2.60)	8.54(3.80)	5.67(3.19)	6.28(2.55)	4.63(1.45)
Nitrogen dioxide	0.0358(0.0142)	0.0392(0.0156)	0.0318(0.0109)	0.0481(0.0142)	0.0234(0.0329)
Sulfur oxide	0.0126(0.0115)	0.0812(0.0665)	0.0175(0.0141)	0.118(0.0973)	0.0284(0.0271)
Oxidant	0.▨▨3(0.0137)	0.0227(0.0128)	0.0229(0.0103)	0.0238(0.0167)	0.0275(0.0099)

Table 2. Mean (standard deviation) – 1968 CAMP data. Ln normal assumption.

	Denver	Philadelphia	Cincinnati	Chicago	St. Louis
	1.575(0.449)	▨(0.454)	1.650(0.382)	1.714(0.810)	1.478(0.344)
		▨6)	−3.504(0.332)	−3.077(0.299)	−3.899(0.657)
			−4.582(1.386)	−2.520(0.951)	−4.195(1.458)
				−4.35▨(▨26)	−3.913(0.535)

measurements are assumed to have a normal and ln normal distribution, respectively. Data were taken from the 1968 Continuous Air Monitoring Program.[1] The ln normal distribution is often used to describe air pollutant variations, and we are using it for evaluation of various indexing schemes. However, we have tested the ln normal distribution and found that it may not always be valid. In Table 3, for example, we show the results of testing the 1968 CAMP data that are summarized in Tables 1 and 2. Calculated statistics are tabulated, with the corresponding (99%) critical values shown in parentheses. We would expect the statistic would exceed the critical value only 1% of the time if the ln normal assumption were strictly valid. It is evident that the assumption is questionable.

Table 3. Kolmogorov-Smirnov test for normality. Ln normal assumption.

	Denver	Philadelphia	Cincinnati	Chicago	St. Louis
Carbon monoxide	0.0844(0.0604)	0.0615(0.0760)	0.1167(0.0668)	0.1955(0.0643)	0.0555(0.0564)
Nitrogen dioxide	0.0615(0.0594)	0.0861(0.0578)	0.0425(0.0595)	0.0393(0.0574)	0.154(0.0562)
Sulfur oxide	0.1456(0.0660)	0.0485(0.0556)	0.1446(0.0600)	0.0644(0.0583)	0.123(0.0552)
Oxidant	0.0927(0.0702)	0.1452(0.0746)	0.0884(0.0961)	0.257(0.0661)	0.118(0.069)

Table 4. Kolmogorov-Smirnov test for normality (Chicago, 1968).

Pollutant	Statistic under		Critical value
	Normality assumption	Ln-normality assumption	
Carbon monoxide	.08097	.19546	.06431
Nitrogen dioxide	.07014	.03925	.05737
Sulfur oxides	.14410	.06443	.05828
Oxidant	.07727	.25698	.0661

Even when separate distributions are computed for a given pollutant on a semi-annual basis — to allow for seasonal variations — the ln normal assumption seems questionable. In Tables 4 and 5, we present results of normality tests for four pollutants in a single city, Chicago, in 1968. The values in Table 4 refer to the entire year. Those in Table 5 were computed on a semi-annual basis: one "season" extending from mid-April to mid-October and the other season extending from mid-October to mid-April. Computation on a seasonal basis appears to follow the ln normal distribution better than does computation on an annual basis, but the assumption remains questionable. The ln normal distribution does, however, seem superior to the normal distribution; hence we are using the ln normal assumption for developing indexing schemes. Clearly, a search for a better distribution is warranted.

Table 5. Kolmogorov-Smirnov test for normality (Chicago, 1968).

| Season | Pollutant | Statistic under | | Critical value |
		Normality assumption	Ln-normality assumption	
1	Carbon monoxide	.07854	.03828	.0800
	Nitrogen dioxide	.06364	.05537	.0825
	Sulfur oxides	.09300	.08491	.0831
	Oxidants	.12203	.254670	.1001
2	Carbon monoxide	.08182	.24803	.1081
	Nitrogen dioxide	.05894	.05443	.0798
	Sulfur oxide	.09544	.07733	.0817
	Oxidants	.12917	.07527	.0881

WEIGHTING FACTORS

When developing any sort of air quality index, one has to be concerned not only with standards and distributions but also with matters such as scaling and weighting. We make a distinction between *scaling* factors and *weighting* factors. Scaling factors operate on observed concentrations of different pollutants measured in various units, make them dimensionless, and put them on a uniform scale — they normalize the different pollutants. Weighting factors, on the other hand, represent an attempt to put a degree of importance on each of the measured pollutants. We do not believe that weighting factors should be the same everywhere. Rather they should reflect the importance of a given pollutant in a given region. This may be a city, county, state, or even a country. Hence if the average SO_2 level is very high in a given city, then we believe that SO_2 is more important there than it might be in another city where the average SO_2 level is very low. Weighting factors should also take account of the normal fluctuations in pollutant levels. They should be dimensionless, and they should also be non-linear, that is, they should increase rapidly as the average level of the given pollutant approaches its standard level. An intolerable situation exists when the average concentration of a pollutant attains or exceeds the standard value.

We define the weighting factor as follows: $W_i = \sigma_i/(S_i - \mu_i)$, where S_i, μ_i, and σ_i are the standard value, mean value, and standard deviation, respectively, for the assumed distribution of the ith pollutant. This equation says that the weight or importance of a given pollutant ranges from σ_i/S_i to infinity as μ_i goes from zero to the standard value. It goes up very rapidly as the average concentration approaches the standard value. It has the property of going negative if the average value exceeds the standard value. When this happens, and it is conceivable that it might, it signals that the standard value needs to be re-evaluated. It says that we expect a hazardous state to exist at least 50% of the time. The importance of that pollutant is excessive, and some action should be taken either to lower the mean value of the pollutant or to modify the standard value if it is not a valid one.

Now what is the significance of $\sigma_i/(S_i - \mu_i)$? This in itself reflects the importance one wants to associate with a given pollutant whose concentration (or some function of the concentration such as the logarithm of the concentration) may be described by a normal distribution with mean μ_i and standard deviation σ_i. In fact, under the assumption that observed concentrations are normally distributed, $(S_i - \mu_i)/\sigma_i$ may be used to calculate the probability that a given pollutant will be less than or equal to the standard. Under the assumption that μ_i and σ_i describe a normal distribution for the observed daily average concentration of the ith pollutant, one would expect 84% of these daily averages to be less than $\mu_i + \sigma_i$ and about 98% of these daily averages to be less than $\mu_i + 2\sigma_i$. Therefore $(S_i - \mu_i)/\sigma_i$ suggests a way to compare and/or perhaps set standards. If the standard is set at $\mu_i + 2\sigma_i$, this is tantamount to accepting an excessive level of the given pollutant 20 days out of 1000 days, over the long term.

Another weighting system could be defined as $W^B_i = \sigma_i/(S_i - B_i)$, where B_i represents a baseline value for the ith pollutant. This weighting system says in effect that the importance of a given pollutant is reflected in the difference between the standard value and its baseline value, *i.e.*, how far the prevailing standard is set from the pristine state. The problem in this sort of weighting system is two-fold. First, it does not take account of the prevailing conditions, only what might be. Second, the determination of baseline values (especially in retrospect) is an extremely difficult and perhaps even impossible task.

Table 6 presents the weighting factors, W_i, for the same four major pollutants and five cities that were cited earlier. The following standard values were used in calculating these factors: for CO, 9.0 ppm; for nitrogen dioxide, 0.05 ppm; for sulfur oxides, 0.14 ppm; and for oxidants, 0.08 ppm. These will be recognized as the current secondary standard concentrations recommended by EPA. Particulate matter is not included because values of this pollutant are not available in the 1968 CAMP data. Hydrocarbon values are not included because the current standard is not consistent with the measurements taken in 1968. It is

Table 6. Weighting factors for ln-normal distribution (1968 CAMP data).

	Denver	Philadelphia	Cincinnati	Chicago	St. Louis
Carbon monoxide	0.722	3.027	0.698	1.674	0.478
Nitrogen dioxide	0.940	1.534	0.653	3.665	0.728
Sulfur oxides	0.527	0.965	0.530	1.718	0.654
Oxidants	0.482	0.556	0.395	0.942	0.386

to be emphasized that the data in Table 6 were computed under the ln normal assumption. This means that μ_i and σ_i refer to the distribution of $\log_e C_i$, where C_i is the 24-hour average concentration of pollutant i.

Note that the weighting factors vary considerably within and between cities. Among the four pollutants, carbon monoxide is given greatest weight in Philadelphia and Cincinnati. Nitrogen dioxide has the greatest weight in Denver, Chicago, and St. Louis. Oxidant is the least important pollutant in all five cities, but it is considerably more important in Chicago than in the other cities.

SCALING OF INDIVIDUAL POLLUTANT VALUES

In our opinion, a desirable scale for an individual pollutant index ranges between 0 and 1. We have adopted the probability that the pollutant concentration is less than or equal to the observed pollutant concentration as the value of the individual pollutant index. This provides a scale ranging from 0 to 1, neither of which is ever attained. With the average value falling at 0.5, the scale changes fastest near the mean. Unfortunately it changes slowly at extremely high values; this is a flaw in the scaling system. The nice thing about this approach to scaling is that the individual pollutant index is the probability that the pollutant concentration will be less than or equal to an observed value. It makes no difference what the prevailing or average condition is; the importance of prevailing concentration appears in the weights, not in the individual scaled value for the day. In other words, the average level of a given pollutant in a city like Philadelphia, for example, might constitute a hazardous condition, and yet the individual index rating for that pollutant would be 0.5 until it is weighted properly.

COMBINING INDIVIDUAL INDICES

We have considered a number of ways to combine the information, $i.e.$, to apply weighting factors to individual pollutant indices and then combine them into a single value. None of these methods is completely satisfactory. In the first place, the composited or overall index value might be less than the highest individual index value. Indeed, one indexing system might consist of the highest individual weighted value in the set. The assumption in that approach would be that if weights are applied to the individual pollutants, they become equal in importance. As scientists, we find it difficult to forego valid data, $i.e.$, to measure many things and then only use one. The usual approach to combining indexing data is to do some sort of arithmetic averaging. This means that high values are decreased and low values are increased. Here then is an expression for an index; it is the most popular one:

$$I_a = \frac{\sum_{i=1}^{k} W_i P_i}{\sum_{i=1}^{k} W_i}$$

in which W_i is the weighting factor for pollutant i, P_i is the individual index value for pollutant i, and there are k pollutants.

The real disadvantage that we see in this is that an excessive value — one that is above the standard — may get averaged down to such an extent that no indication is given of a bad situation when one exists. The overall index rating would indicate acceptability when in fact some pollutants exceed their standard value. Some signal is needed to indicate when any pollutant exceeds its standard value. Negative W_i values are not allowed in this system.

Another disadvantage to this approach is that pollutants with large weighting factors can dominate the index value through the ΣW_i term. Moreover, the index value is dependent upon *which* pollutants are measured, again through the ΣW_i term. For these reasons, we prefer to index by the following expression:

$$I_1 = \frac{\sum\limits_{i=1}^{k} W_i P_i}{k}$$

This system is not influenced unfairly by the number of pollutants involved or by their identity.

Geometric averaging has not been used for constructing air pollution indices, to our knowledge. The geometric average is given by

$$\left[\prod_{i=1}^{k} P_i \right]^{1/k}$$

We have been particularly interested in this approach because the term within the brackets constitutes the *joint probability* for k independent random events. Hence, if k pollutants are measured and scaled according to the probability of occurrence, and if they are independent random occurrences, then the term within the brackets corresponds to their joint probability. Unfortunately, as numbers smaller than unity are multiplied, the product decreases; so the joint probability becomes very small. It must be scaled back up into a meaningful or understandable value. For indexing purposes, it also should allow different numbers of pollutants to be measured. The geometric mean, with its $1/k$ exponent, thus seems attractive as a scaled joint probability for k pollutants. Unfortunately, all pollutants have equal weights at this point: the weighting factors are $1/k$ for each pollutant and they sum to unity. This is a characteristic of the geometric mean.

To scale and weight the individual pollutant indices and to allow k to vary, we prefer to weight terms exponentially and to then take the kth root of the product:

$$I_2 = \left[\prod_{i=1}^{k} P_i^{(W_{min}/W_i)} \right]^{1/k}$$

where W_{min} refers to the smallest W_i among the k pollutants that enter into the computation. Using the ratio of W_{min} to W_i insures that the exponent is always 1 or less, which is necessary because the P_i values themselves range between 0 and 1. Placing W_i in the denominator of the exponent imposes increasing values on the terms as their respective W_i's increase. Moreover, the overall index now ranges between 0 and 1 and is applicable to a variable number of pollutants. Its terms are individual probabilities, weighted according to the identities of the k pollutants that enter into the computation. The flavor of statistical validity is retained.

In Fig. 1, I_1 and I_2 values for Philadelphia are shown, based again on 1968 CAMP data. Our computer program calculates weighting factors, individual index values, and the overall indices, and then prints and plots the information. These curves are representative output plots. The number of pollutants involved in each computation is indicated by the key symbols (0, Δ, +, X) that mark the overall index value for a given day. We have similar plots for the other cities. The point to be made from this figure is that the index systems vary considerably on a day to day basis – they respond to changing conditions. Note that the I_1 values do not have fixed range, but the I_2 values fall within the range 0 to 1.

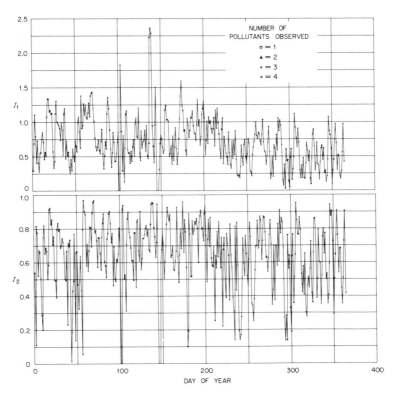

Fig. 1. Arithmetic and geometric index values for Philadelphia (1968 CAMP data).

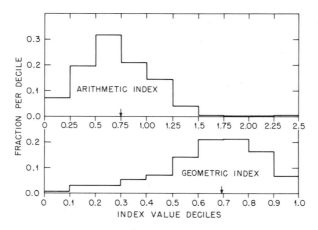

Fig. 2. Histograms of index values for Philadelphia (1968 CAMP data).

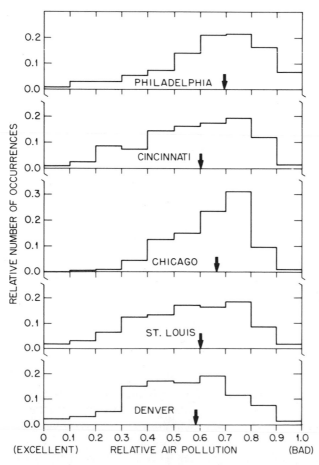

Fig. 3. Histograms of geometric index values for five case cities (1968 CAMP data).

The responses of these indices are reflected better by the histograms shown in Fig. 2. The horizontal axes contain the range of index values for 1968, but the scale is divided into ten parts of equal length. The vertical axes indicate the proportion of the index values that fall within a given index range. Thus approximately 42% of geometric index values for Philadelphia had a value between 0.6 and 0.8 during 1968. The arrow at 0.76 indicates the arithmetic index value that corresponds to all pollutants existing at the median value, *i.e.*, individual index values of 0.5. Similarly, the arrow at 0.692 on the lower histogram corresponds to average conditions in Philadelphia. These values for I_1 and I_2 reflect the influence of Philadelphia weighting factors, of course.

It is also interesting to compare histograms for several cities during the same time period. In Fig. 3, histograms for the five case cities are shown. The axes are analogous to those of Fig. 2. Again, indices corresponding to average conditions are indicated by arrows. From these curves, we note that the air pollution in St. Louis and Cincinnati, as reflected by the geometric index, varied considerably during 1968, and the pollution levels in these two cities were comparable. By contrast, the variation in air pollution in Chicago and Philadelphia was less, but the levels were higher. Denver's air pollution level was lower than that of the other cities, and it varied an intermediate amount.

FUTURE WORK

All of the information presented here has been based upon 1968 CAMP data. These indexing systems need to be observed over a period of time, using current recommended measurement techniques and standards. Comparisons of the response of these and other indexing schemes need to be made, both among themselves and with observed pollution levels. We are currently collaborating with the Knox County Air Pollution Control Board, Knoxville, Tennessee, in collecting the information needed for these comparisons.

We are planning to look at other CAMP data in some detail to see just what averaging times are appropriate, to see if the ln normal distribution is adequate or if another distribution is preferable, and to devise ways to incorporate synergistic effects into the index. We want to study "effects statistics" to determine if compatibility exists between them and these indexing systems.

CONCLUSION

Because of the current information explosion, indexing work seems to be much broader in context than we have implied. In many areas of science, we need to be considering ways to combine information into a minimum number of meaningful and understandable terms — to simplify data as much as possible. This type of activity requires multidisciplinary action and offers a means for increasing multidisciplinary communication. The overriding criterion is to preserve validity. Hence, while we examine the question of indexing air quality, we are searching for other significant areas that might benefit from indexing work. In time, indices are likely to be used by scientists in many fields.

REFERENCES

1. Air Quality Data, 1968 Ed., National Air Pollution Control Administration, U. S. Department of Health, Education, and Welfare, Washington, D. C., Nov. 1969.
2. Liliefors, H. W. *J. Am. Statistical Assoc.*, 62: 399–402 (1967).

INDICATORS OF ENVIRONMENTAL NOISE

David M. Lipscomb

*Associate Professor of Audiology and Speech Pathology and
Director, Noise Study Laboratory
University of Tennessee
Knoxville, Tennessee 37916*

"Make a joyful noise . . . " Psalms 100:1
"Be still and know . . . " Psalms 46:10

Noise in the environment is not an altogether new area for consideration. As noted above, scriptural references are available that provide insight into the presence of noise as a topic of note several hundred years ago. In addition, great world literature has known occasional references to annoying sound levels such as the invective of the Roman poet Horace (Epist. ii. 2) against "the squealing of the filthy sow and the barking of the mad bitch" — two environmental sounds that he would like to have eliminated. The 18th century philosopher Schopenhauer castigated the "infernal cracking of whips in the narrow resounding streets of the town . . . " and went on to suggest that such sounds, " . . . must be denounced as the most unwarrantable and disgraceful of all noises."[95] Thus appeared an early denunciation of traffic noise.

It is indeed humbling to be asked to summarize the study of noise, for there are many remarkable persons whose labors have allowed us to determine some of the features of noise exposure. In such a presentation as this, it seems most appropriate to single out a small number of those whose efforts have touched the professional life of many of us who are relative newcomers to the field. In 1119, Bernard of Chartres[8] wrote:

"We are like dwarfs seated on the shoulders of giants.
    If we see more and further than they, it is not
due to our own clear eyes or tall bodies, but
    because we are raised on high and upborne by their
gigantic bigness."

One of the earliest voices to express concern about the modern technological development as producing an onrush of noisy products was that of Dr. Vern O. Knudsen.[60,61] Following shortly thereafter with "gigantic bigness" were such eminent personalities as Aram Glorig,[46] Karl Kryter,[65] S. S. Stevens,[116-118] E. Zwicker,[136] W. Dixon Ward,[126-128] Cyril Harris,[51] and Leo Baranek.[16] The work of these men and the countless others who have not been specifically cited is both impressive and voluminous.

THE NOISE PROBLEM

A common catch phrase in the popular literature alludes to "the noise problem." It is important to note, however, that the subject has not yet been reduced to a definable entity. It is safe to estimate that none of the workers in this area of professional concentration understand the many factors sufficiently to be able to encapsulate it with a single encompassing description; thus, the problem of noise remains largely a mystery.[129] This is because we do not yet understand the parameters nor the depth and breadth of each enough to comprehend the full effect of sound stimulation upon human (and animal) existence.

In addition to the above problem of definition, one must consider the wide range of individual differences in responsiveness to various sound stimuli. It can be said, for example, that one man's noise is his son's music. Mental set, orientation, experience, personality, health, and a myriad of other personal factors confound the attempt to establish simplistic indicators with our present knowledge.

SUGGESTED INDICATORS

There is a sizable accumulation of knowledge in certain areas, which provides an understanding of the quality of our acoustic environment. It must be emphasized that the space provided for the discussion of these indicators is understandably severely restricted; thus, each is presented in a somewhat abbreviated form.

Physical Indicators

Acoustic instruments have been utilized for many years to quantify the physical properties of sound in order to classify certain acoustic events according to such major parameters as amplitude, duration, and spectrum. General purpose sound-level meters are used to measure the sound present at a given instant or during a given set of event conditions.[3,4] These devices consist of a good quality microphone, an amplifying circuit, an indicating meter, and a power supply. The circuitry can usually be modified by manipulating a switch so that the frequency response of the circuit is altered to meet one of three standard weighting scales (Fig. 1). The C weighting scale allows a flat response to frequencies between 50 and 5000 Hz with slight rolloff above and below those frequencies. The A scale is thought to approximate the frequency response of human ears by slightly attenuating those frequencies below 1000 Hz. The infrequently used B scale has a frequency response that lies in the midrange between A and C.[55] In recent years, the A scale has found favor as the measuring mode for determining human correlates of sound. It must be remembered, however, that the A scale is only an approximation of the frequency response of the human ear, and A-scale readings must be interpreted in that light.

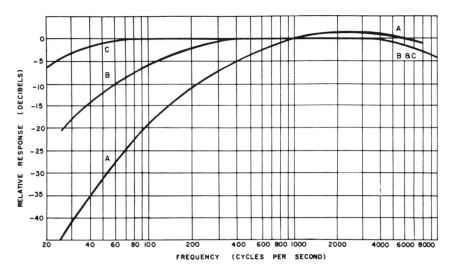

Fig. 1. Frequency response characteristic of standard weighting scales for sound-level meters.

As a part of the burgeoning of the instrumentation field, more sophisticated technology has provided numerous additional means for the measurement and analysis of sound. The octave band analyzer[5],[6] is designed to utilize precisely tuned filters in order to sample octave elements of a sound spectrum around a standardized center frequency. The preferred center frequencies range in octave intervals between 16 Hz and 16 kHz. For more definition of spectral content, $\frac{1}{2}$, $\frac{1}{3}$ and $\frac{1}{10}$ octave analyzers are currently available to sample the 16 Hz to 16 kHz frequency range. Computer technology has further refined the analytical capability of instrumentation so that current versions of spectrum analyzers have the capability of measuring the acoustic energy present in frequency bands less than 1 Hz wide.

It is well established that physical measures of sound in the environment can be made with exceptionally good reliability and accuracy.[98] Recent additions to the acoustician's armamentarium have increased his ability to overcome the transient and unpredictable signature of sound. Yet a most important factor cannot be overlooked. The receptor of the sound is man. Herein, psychoacoustic principles come into play.

All men do not respond to sound in the same way. There is a wide range of human sensitivity to sound; therefore, some perceive sound as being more intense than do others. Numerous psychological reasons account for additional variation between persons in their responsiveness to sound stimulation.

Attempts to quantify the perceptions of loudness growth have met with some degree of success. Several scaling methods have been developed and have served as the springboards for further scale development:[89]

Phon scale. The phon scale, proposed by Fletcher and Munson[44] in 1933, used the concept of equal loudness, loudness level, or equivalent loudness.

Scaling was accomplished by comparing test tones to a reference tone of 1000 Hz. The sound pressure level (SPL) of the reference tone (in dB) was then the value given to the test tone in phons. For example, a 125-Hz tone adjusted in amplitude to sound equally loud as a 1000-Hz tone of 40 dB (SPL) is given the value of 40 phons. This scale has also been used, with limited success, in determining the "loudness" characteristics of complex noise.

Sone scale. The sone scale was generated by asking listeners to judge when one sound was perceived to be "twice as loud" or "half as loud" as a reference sound. Although this sounds like a most difficult undertaking, it was discovered that trained listeners could make these judgments with a good deal of consistency. The unit for this loudness scaling method is the *sone*. The value of one sone was arbitrarily established to be 40 phons, that is, the loudness experienced by a listener when stimulated with a 1000-Hz tone whose sound pressure is 40 dB. The amplitude setting of the sound whose value is 2 sones would be that adjudged by the listener to be twice as loud. A general rule has been given as: each increase of a sound by 10 phons will double the loudness value in sones.[24]

Noy scale. It became apparent after an attempt to equate sound level measures of aircraft flyovers with noisiness judgments that the sone scaling did not hold at all levels. Kryter and Pearsons[66] then established the *noy* scale by using the same technique as for the sone scale with the exception that the listener was asked to judge "noisiness" rather than "loudness." This introduced into the scale the highly subjective, but necessary, "irritation factor." This scale was revised[64] later to incorporate the octave band concept, providing the foundation for the next scale to be discussed.

Perceived noise level. Kryter observed that the high-pitched whine of the jet engine was seemingly more annoying than was the sound of the propeller motor. He, therefore, was moved to design a scale that would give more weight to the higher frequencies in the aircraft sound spectrum.[63] This scale, called *perceived noise level* (PNdB) is calculated with the use of the noy scale and provides a single-number value of the noisiness of aircraft flyovers. Although this scaling method and modifications of this technique (such as Effective Perceived Noise Level — EPNdB[65]) are in considerable use for the determination of aircraft noise levels, they have not found much favor in other areas of environmental noise study.

Other scales. The frequency of occurrence of a given acoustic event has a great deal to do with the acceptability or non-acceptability of that sound source. It is deemed necessary, therefore, to count the number of occurrences per unit of time in order to project the irritating quality of a given sound. The British Noise and Number Index (NNI) and the *German Q Index* are proposed for this purpose.[24] The Noise Exposure Forecast (NEF) can be calculated in order to project the zones of irritation around a particular noise region.[17] This scaling device is especially popular in relating the future implications of airport development. A NEF of 30 or greater is considered to be unsatisfactory for residential neighborhoods.

Robinson[102] has described a method whereby the Noise Pollution Level (NPL) is calculated by totaling the "energy mean" of various noise levels for specific periods of time. These figures are compounded by other calculations that are generated to provide a single-number sampling of the dissatisfaction particular noise sources bring the person or persons experiencing the sound.

Another approach which has some very good points is advanced by Goldstein.[49] His calculation of a noise index is part of a large project on environmental monitoring under the sponsorship of the Council on Environmental Quality. This method accounts for the otohazardous aspects of intense sound. In addition, irritating sounds that may not endanger the ear but which contribute to the overall noise-related anxiety level are also considered. The basis for Goldstein's work is beyond the establishment of auditory damage-risk criteria and " . . . is intended to portray environmental quality." His method utilizes a distribution of noise exposures which might be typical for a 24-hour period. The measured sound environment is then related to the standard proposed by Goldstein, and an index of the quality of the acoustic environment is obtained.

The problem with using purely physical measures of sound as indicators of environmental quality is that similar acoustic events arouse widely differing reactions among individuals.[90] With the use of psychoacoustic techniques, there are still sufficiently great variations that the physically based single-number environmental quality index is still a gleam in the psychoacoustician's eyes.

To further confound the situation, there is a trading relationship between the amplitude of a signal and its duration. This relationship is affected by the spectrum of the sound and by the frequency of occurrence.

Physiologic Indicators

Noise has long been known as an irritant. Operationally defined as unwanted sound,[51,123] noise gives rise to a multitude of physiologic reactions which have been listed by a host of researchers.[87,131] The pattern of physiologic reactions varies within and between individuals.

As yet, the basic question, "Is noise a stressor?", has not been fully answered. Some investigators have sought to observe certain forms of stress reaction in experimental animals by histochemical analysis of certain portions of the system. An appropriate area to study is the adrenal cortex. In Argentina, Arguelles et al.[10,11] have indicated that noise stimulation results in an increase in adrenal cortex activity in human subjects. They reached this conclusion: "Noise proved to be an important stressing agent capable of causing disturbances in cardiovascular and psychotic subjects."

Recognizing the brilliant work on stress by Dr. Hans Selye[108-110] and his colleagues, we chose to test the question by exposing a group of rats to noise and observing the classic stress reaction triad first described by Selye as (1) atrophy of the thymus gland, (2) development of duodenal ulcers, and (3) swelling and discoloration of the adrenal glands.

Rats were placed in a sound treated enclosure into which was presented a broad-band white noise adjusted to a level of 110 decibels (dB), sound pressure level (SPL). The animals were left in the noise for a continuous period of 48 hours after which they were terminated still in the presence of noise in order to eliminate the possibility of spontaneous recovery upon being removed from the sound. Control rats were placed in a quiet enclosure for the same period.

Examination of the internal organ systems of the noise-exposed rats yielded convincing evidence, in that every noise-exposed rat demonstrated at least one of the features in the triad of stress reaction. Most of the animals were found to manifest all three forms of the reaction. Noise-stimulated animals exhibited either atrophied or severely discolored thymus glands. There were duodenal ulcers in over half of the rats. The adrenal glands were commonly noted to be enlarged, and the sheath of fat which normally surrounds the adrenals had been noticeably reduced. The control animals removed from the quiet enclosure showed significantly fewer of these symptoms.

We have sufficient data on two experiments to report here. A group of adult rats placed in either noise or quiet yielded the results cited in Table 1, in which

Table 1. Results of the examination for abdominal signs of stress reaction for three groups of adult rats.

Experiment	Animal	Thymus	Duodenum	Adrenals	Stress indicator ratio
Group 1 (Noise)	R1	+	+	+/+	
N=13	R2	+	+	0/+	
	R15	+	+	+/+	
	R16	+	+	+/+	
	R17	+	0	+/0	
	R18	+	+	0/0	40 of 52
	R21	+	0	+/+	(76.9%)
	R22	+	+	+/+	
	R23	+	+	+/+	
	R24	+	0	+/+	
	R25	0	+	+/+	
	R26	+	0	0/0	
	R27	+	0	+/+	
Group 2 (Quiet)	R3	+	0	0/0	
N=4	R4	0	0	0/0	1 of 16
	R12	0	0	0/0	(6.3%)
	R20	0	0	0/0	
Group 3 (Cage)	R5	+	0	0/0	
N=5	R6	0	0	0/0	
	R7	+	0	0/0	2 of 20
	R13	0	0	0/0	(10.0%)
	R14	0	0	0/0	

+ = positive indication of stress reaction
0 = no indication of stress reaction

a plus sign (+) signifies an observable deviation of the structure. These deviations were designated as stress signs. There was, according to our classification, one stress sign for the thymus and one for the duodenum plus two possible stress signs for the adrenals (discoloration and deterioration of the fatty sheath). Thus, the animals were rated on the basis of four potential stress reaction signs apiece.

Of the 13 adult rats, only one demonstrated less than two stress signs, presumably as a function of intense noise stimulation for a continuous 48-hour period. The group was found to have 40 observable stress signs from a possible number of 52. This is computed to be a stress indicator ratio of 76.9%.

Four rats were placed in a quiet chamber for the same period of time so that their experience was the same as that of the noise-stimulated rats with the single exception being the sound environment. One of these animals was found to have a positive stress sign (thymus atrophy) yielding a stress indicator ratio of one out of 16 (6.3%).

A third group of adult rats were taken from their cage and prepared as described above. Two of these animals were found to have slight deviations from normal thymus glands. The stress indicator ratio for the cage controls was two for 20 (10.0%).

The results of the first experiment lead to the conclusion that noise produces the triad stress reaction with a high degree of consistency when noise-stimulated animals are compared to non-noise-stimulated ones.

The presence of some thymus gland deficiencies in the control populations was felt to be a spontaneous defect as a result of some unknown cause. To test this concept, young rats were used in a second investigation.

The results of this experiment can be seen in Table 2. The young rats (21 days old) that were placed in noise for 48 hours demonstrated a stress indicator ratio of 11 of 20 possible signs (55.0%). One of the rats confined in a quiet enclosure demonstrated two stress signs, and two others showed one sign apiece, giving a ratio of 20.0% for the second group. The five cage control rats were found to have no positive stress signs.

These animals had just been weaned; thus, the positive stress signs in the quiet control group might be understood as a combination of the weaning adjustment and the stress of being confined during the experiment.

Additional observations in some noise-exposed animals included a general constriction of the intestinal tract and occasional ulceration of portions of the intestinal tract other than the duodenum; one animal appeared to have suffered a coronary thrombosis.

It was rather startling to note that these conditions could be seen in rats after only 48 hours of exposure to a noise not unlike that experienced by many industrial workers. We concluded that noise is, in the sense of a classical definition, a stressor.

Elsewhere, significant work is currently underway to bring to light other physiological effects of being exposed to irritating sound sources. The term "irritating" is used here purposely to indicate that the body, according to some

**Table 2. Results of the examination for abdominal signs of stress reaction
for three groups of baby rats.**

Experiment	Animal	Thymus	Duodenum	Adrenals	Stress indicator ratio
Group 1 (Noise)	R29	0	0	+/+	
N=5	R30	+	+	+/+	
	R31	+	0	0/0	11 of 20
	R32	+	+	0/+	(55.0%)
	R33	0	0	+/0	
Group 2 (Quiet)	R34	0	0	0/+	
N=5	R35	0	0	0/0	4 of 20
	R36	+	0	0/+	(20.0%)
	R37	0	0	0/0	
	R38	+	0	0/0	
Group 3 (Cage)	R39	0	0	0/0	
N=5	R40	0	0	0/0	0 of 20
	R41	0	0	0/0	(0.0%)
	R42	0	0	0/0	
	R43	0	0	0/0	

+ = positive indication of stress reaction
0 = no indication of stress reaction

students of the subject,[37] attempts to regain a condition of equilibrium after having been unbalanced by its reaction to the irritant (in our interest, sound).[12] During this period, the "counterrelevant forces" within the body vie for control, altering the emotions, general health, and ability to perform mental or motor activities with ease and accuracy.

One attempt to describe the alterations in body function is the N-response.[37] This syndrome includes: vasoconstriction and rising blood pressure with slight elevation in heart rate; slow, deep breathing; measurable change in skin resistance to electricity; and a variation in skeletal-muscle tension.

Other researchers have confirmed these observations[23] and have also suggested that digestive system changes occur,[36,115] glandular activity alters the chemical content of blood and urine,[50,74] vestibular problems occur,[39] pupils of the eye dilate,[57] fetal development may be disturbed,[113] and sleep is interruped.[122] There is evidence that sound pulses can modify cardiac rhythm.[34] Some researchers have produced fatal audiogenic seizures in laboratory rats with the use of sudden intense sounds.[53] These factors, independently or in combination, contribute to feelings of fatigue, irritability, or tension.

A word of caution is necessary. It is very easy to project the results of these studies into a "doomsday" prediction. Unfortunately, some people are not resisting the temptation to do so. We must be reminded, however, that the greatest bulk of this research has been conducted with nonhuman subjects, and therefore, the projection to human reaction cannot be easily made. The most appropriate interpretation of the data is to realize that inordinately great exposure to noise has a potentially deleterious effect upon the vital physiologic processes and must be avoided if one is to remain free of the types of disturbance such exposure might cause.

Some would state the interpretation even more cautiously, for they hold that the weight of evidence is equivocal. Long-term studies will ultimately determine whether the predicted devastating side effects of excessive noise exposure and the resulting reaction to the noise are viable. To date, there is scant evidence on either side of the argument, for man has never before been forced to endure an acoustic environment composed of such high level sounds as in this age; therefore, his response to the sound is not fully predictable.

There is, however, some interesting related data on aging which was obtained by studying a remote primitive tribe, the Mabaans, in the Sudan.[103,104] Not discovered until 1956, the Mabaan tribe was found to be functioning at about the late stone age level of civilization. Their bland, monotonous diet coupled with an extremely noise free manner of living was regarded by a team of medical researchers as a significant departure from the life style of "civilized" man in the industrial nations. Extensive physical examinations and hearing evaluations of the Mabaan tribesmen of ages ranging from 10 years to over 80 years revealed an inordinately low incidence of upper respiratory defects, few cases of cardio-vascular diseases, virtually no disorders of the intestinal tract, and near normal hearing even in the aged members of the tribe. It was pointed out by the research team that the hearing of 80-year old Mabaan tribesmen was better than the average 30-year old American.[105]

Although the results of the Mabaan and similar investigations must necessarily be interpreted with caution,[15,106] there is strong evidence that the aging processes as well as the occurrence of internal disorders might be at least augmented by the presence of high environmental noise.[107]

Auditory and Vestibular Indicators

Hearing threshold shifts. High level noise has a noticeable and measurable effect upon the auditory and vestibular end organs. The existence of temporary threshold shifts (TTS) after exposure to loud sounds can be seen by testing the hearing of persons prior to and after exposure.[2,65,126] The amount of reduction in hearing sensitivity (threshold) is determined as the TTS. Any sound capable of causing a TTS of 30 dB or greater is generally regarded as being dangerously high in intensity, and exposure to such sounds should be avoided.

The above description must be restricted to sounds that are continuous or have a "steady state" quality. Those sounds whose amplitude is great, but which occur for very brief durations, are frequently described as "impact" or "impulse" sounds and are to be considered in a class by themselves with respect to their TTS features.[29,65,126-128]

Continued stimulation by dangerously high sound levels will most likely result in a permanent threshold shift (PTS). Whereas a TTS is reversible when the person is removed from the noise, a PTS is irreversible and connotes damage to the sensory cells in the inner ear. Once these cells are destroyed, there is no known mechanism for their regeneration, so that repeated exposure to high level sounds will result in an accumulation of destruction in the sensory cell population, increasing the breadth of effect. The first measurable indication of such an occurrence is the persistent reduction in hearing acuity for the audiometric high frequencies (2 kHz to 8 kHz). As the PTS becomes greater in magnitude, the loss of hearing acuity for some of the lower frequencies may become evident.

Damage Risk Criteria (DRC) have been developed by determining the amount of TTS caused by sounds of known amplitude and duration.[1,25,48,67,124,125,130] These criteria serve as indicators of the danger level for noise exposure. If a person's sound stimulation exceeds the DRC, it is quite probable that his hearing mechanism will sustain injury to some extent.[76]

One of the factors of aging includes the decrease in hearing ability for high frequency tones. As age increases, sensitivity for these test signals decreases. It is anticipated, however, that young persons between the pre-teen years and young adulthood should have normal hearing throughout the audiometric test frequency range with the occasional exception of the young person who has sustained a hearing loss for one of several medical or hereditary reasons.[52] One indicator of the existence of inordinately high levels of noise exposure would be to note whether young persons have sustained measurable hearing losses in the high frequencies. The information[88] was summarized and presented to a Congressional Subcommittee in June of 1971.

In the spring of 1968, three studies were undertaken in the Knoxville, Tennessee, City Schools.[33,96,135] In the studies, a total of 3000 students at three grade levels were given modified hearing screening tests in order to determine the prevalence of screening failures for the high frequency tones (above 2000 Hz). Of the sixth graders tested, only 3.8% of the students failed the high frequency screening criterion (15 dB, Ref. 56). This figure rose to 11.0% for the ninth grade population and held at approximately the same level for the high school seniors (10.6%).

The apparent trend to greater failure rates in the older students led to a similar hearing survey of college students. In the fall of 1968, a total of 2769 incoming freshmen at the University of Tennessee between the ages of 16 and 21 years were given the same modified screening test used earlier in the public schools. The laboratory staff was concerned to note that 32.9% of the students failed to meet the failure criterion. To confirm that striking finding, a portion of the incoming class (1410 students) was screened for hearing in the fall of 1969. Rather than noting a decrease in the prevalence of high frequency hearing effect, the reverse was true; the survey yielded an incidence indication of 60.7%. These results are illustrated in Fig. 2.

These data offer evidence, based upon measured hearing levels of 7179 young persons age 21 or younger, of a trend toward loss of high frequency

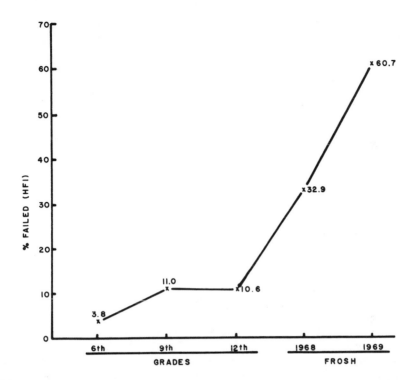

Fig. 2. Results of five studies that assessed the hearing abilities of students. Percentage indicates number of students who failed to hear at least one of the test tones in at least one ear.

hearing acuity of awesome proportions. It must be understood that the great majority of the students did not manifest serious hearing impairments. In fact, most of those who were found to have failed by a slight degree were unaware of any loss of acuity. The point remains, however, that the population from the age range tested should have a considerably smaller number with measurable high frequency impairments than was found.[86]

The audiometric test frequency that was failed with the greatest consistency was a 6000-Hz tone. Much of the previous work in this area has cited the 4000-Hz frequency as being the most convenient indicator of early auditory damage due to noise exposure.[35] Another recent study, however, supports the observation that 6000-Hz hearing impairments are more prevalent than are 4000-Hz losses.[92] The data are incomplete at present; thus, the controversy is unresolved.

Further analysis of the data revealed that the males in the studies demonstrated considerably greater prevalence of hearing effect than did females. This observation is in concert with other research which has preceded these studies.[92,97]

Of course, it is not feasible to attribute the rise in prevalence of high frequency hearing impairment singularly to noise exposure. One can reason, however, that the popularity of high intensity recreational sound sources such as live rock music,[31,38,40,45,72,73,77,82,101,112,133] sport shooting,[121] motorcycling,[85] and sport racing, all coupled with the apparent rise in community noise levels,[28] should be considered to have a potentially distinct effect upon the auditory acuity of young persons.

There is a direct relationship between high level noise exposure and the advent of a ringing sensation originating in the ear. This along with two other factors will serve to introduce subjective indicators of dangerously high environmental noise:[80,81]

1. If the ringing sensation, known as tinnitus, is noted after stimulation by noise, ear damage may have occurred and subsequent exposure is not advisable.
2. When, in the presence of high level sound, voice communication is extremely difficult or impossible, the sound level is dangerously high. A common environment in which this condition persists is the discotheque, which is filled with greatly amplified sounds generated by contemporary musical aggregations.[45,101]
3. The ears have been pushed to, and perhaps past, the danger point if one consistently notices that his hearing sensitivity is lessened upon leaving a noisy environment. Although it is common to recover most of one's hearing after a sufficient rest period in quiet, recurring experiences of TTS will ultimately increase the probability of the threshold shift becoming permanent.

Cochlear cell damage. We have found, with the use of experimental animals, that high intensity sound with energy in a broad range of frequencies is capable of causing widespread destruction of the irreplaceable sensory cells in the cochlea (inner ear).[78,79,84,91] By the use of laboratory techniques, animals (guinea pigs, chinchillas, or rats) have been exposed to high intensity sound. After an appropriate waiting period in order to allow scar tissue to form, the inner ear tissues were dissected out of the temporal bone of the skull and studied with use of a high power light microscope.[19,41,62] In normal cochlear tissue, a very regular, geometric configuration of sensory cells is seen (Fig. 3). In animals that have been exposed to high noise levels, however, the cell patterns are interruped and damage is seen readily (Fig. 4).

Similar experiments have been conducted wherein impulse-type noise was used as the sound source.[93,99] It is important to note that the manifestation of noise damage in these studies should be treated separately, in that impulse noises seem to operate in the ear quite differently than do steady state sounds.

A very interesting and important observation has been made with respect to the pattern and location of broad-band noise-induced cochlear destruction. The injured areas are not restricted solely to the high frequency regions of the inner ear, as one might suspect from the audiometric data presented earlier. In reality,

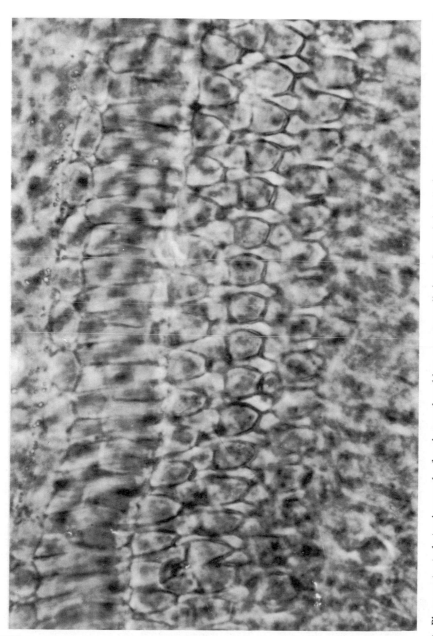

Fig. 3. Phase contrast photomicrograph of undamaged cochlear receptor cell tissue. Inner hair cells form a single row across the top; three rows of outer hair cells can be seen in the the middle of the field (Oc. 10×; Obj. 40×).

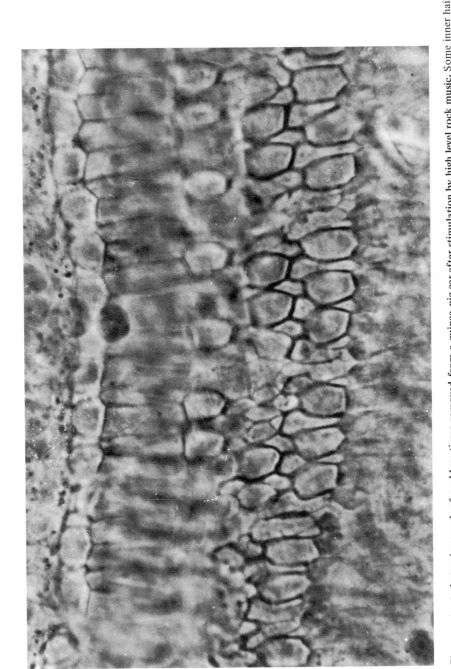

Fig. 4.   Phase contrast photomicrograph of cochlear tissue removed from a guinea pig ear after stimulation by high level rock music. Some inner hair cells are damaged, and extensive damage can be seen in the outer hair cells (Oc. 10X; Obj. 40X).

the damage is consistently greatest in the regions of the inner ear which are generally regarded to be responsible for responding to tones below 2000 Hz.[91] This information points up the possibility that conventional hearing tests do not adequately measure the structural integrity of the portions of the inner ear which deal with low frequency information.

From this data we might speculate that persons who manifest a noise-induced high frequency hearing impairment may have sustained considerably greater inner ear sensory cell destruction than the hearing tests might indicate. In that event, the gravity of the hearing survey discussed earlier is amplified.

The results discussed above relate only to the use of noise sources whose spectrum is wide. Stockwell et al.[120] have used the same laboratory approach in evaluating the cochlear damage occasioned by exposure to high intensity pure tones. They have found relatively restricted areas of damage which relate quite closely to the frequency specificity of the cochlea. As the intensity of the stimulus was increased, however, the breadth of cell destruction widened concurrently.

Other investigations have noted additional types and forms of hair cell manifestations. Bredberg et al.[20] observed a strange modification in the structure and form of stereocilia atop the hair cells. After noise exposure, the cilia were viewed with the use of a scanning electron microscope. Fusion of cilia took place, and there developed giant hairs whose height greatly exceeded that of normal components. Lim and Melnick[75] reported the same views and added that the cuticular plate, the top surface of the hair cell, was distorted and appeared swollen. Further evidence of the destructive aspects of otohazardous noise will be forthcoming as the use of scanning electron microscopy spreads.

By combining the findings of hearing surveys with laboratory data, we have an indicator that provides cause for concern. There is evidence that the hearing acuity of a significant portion of young persons under 21 years of age is becoming reduced many years before such reductions should be expected. These implications lead to the fearful suggestion that the current population of young persons will encounter much more serious hearing problems in their middle years than the present group of 50 to 60 year-olds. Although high level noise must not be singled out as the only factor in this apparent "auditory epidemic," it must be considered to be a significant contributor. There appears to be a sizable degree of auditory deficit in persons of an age group that only a few years ago comprised the subjects for our current audiometric norms.[56]

In fact, it is appropriate to suggest that a normative study for the purpose of establishing or validating audiometric standards is virtually impossible at present in the major industrial areas of the civilized world because of damaged hearing within the population on which norms should be established.[7]

Interruption of cochlear and vestibular blood supply. It was noted earlier that one of the physiologic responses to high noise stimulation was constriction of the veins and arteries. Recent studies have indicated that one result of noise stimulation is a reduction in the blood supply of the inner ear region.

Lawrence[69,70] has demonstrated that the number of red blood cells (erythrocytes) in the inner ear of an experimental animal can be noticeably reduced when noise is presented to the animal. He concludes that the reduction of blood cells in the vicinity of the organ of Corti, the end organ of hearing, may account for TTS.[71]

In some work related to the stress studies mentioned earlier, we have observed the presence of blood cells in the capillaries of the auditory and vestibular end organ regions. Several interesting anomalies have been noted:

a. The number of red blood cells in the capillaries immediately under the auditory sensory cells are considerably reduced in the noise-stimulated animals (see Fig. 5).

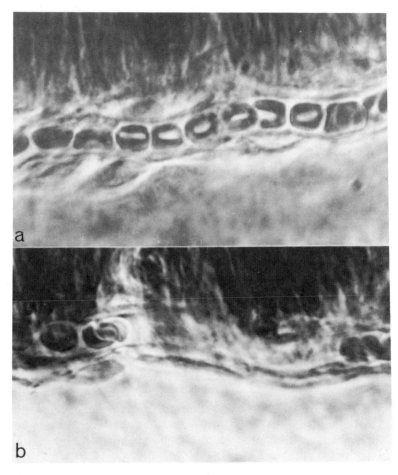

Fig. 5. Comparison of cochlear capillary content of two rats. Top: a capillary filled with erythrocytes taken from a quiet control animal. Bottom: capillary showing reduction in blood cell presence after noise stimulation (Oc. 10×; Obj. 100×, oil).

b. An experiment with guinea pigs resulted in the males showing a somewhat greater reduction in the presence of red blood cells than females although all animals received the same amount of noise. Perhaps this partially explains the observation that females do not suffer as much PTS as males although the stimulation is the same. It is commonly stated that females have "tougher" ears than do males.[86,92,97]

c. It was observed that rats exposed to noise demonstrated a reduction of red blood cells in the capillaries adjacent to the vestibular end organs as well as in the cochlea (see Fig. 6). It is known that one of the common complaints of persons in noise is that they feel a sense of imbalance. They describe experiencing an unsteadiness that may be a vestibular side-effect of the auditory over-stimulation and concomitant reduction in blood supply for the entire stato-acoustic sensory region.

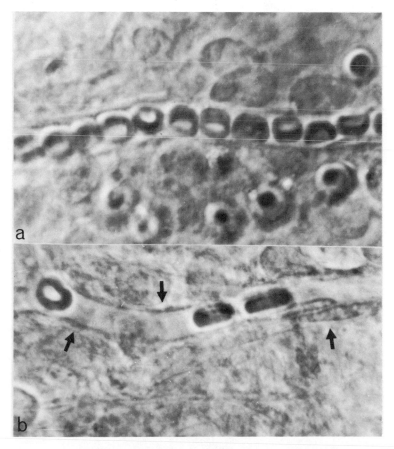

Fig. 6. Comparison of blood cell population in the crista of two rats. Top: vestibular end organ of a quiet control animal. Bottom: same organ in a noise-exposed animal, with dramatic reduction in number of blood cells (Oc. 10×; Obj. 100×, oil).

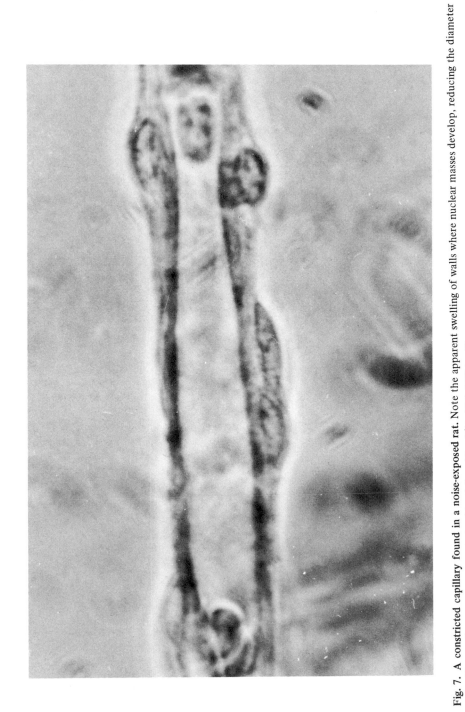

Fig. 7. A constricted capillary found in a noise-exposed rat. Note the apparent swelling of walls where nuclear masses develop, reducing the diameter of the capillary and consequently restricting blood flow (Oc. 15×; Obj. 100×, oil).

The capillaries in the auditory and vestibular areas seem to constrict by developing nuclear masses which thicken the walls of the capillary thereby reducing the internal diameter of the passage. The result is the "choking off" of the capillary, creating sizable gaps between red blood cells and bringing the normal blood flow to a halt in some of the capillaries as shown in Fig. 7. The distribution of oxygen to the sensory cells and their supporting tissues is thus interrupted or rendered less effective.

The mechanism of noise damage. It is not fully known how many mechanisms of noise damage exist. Speculation has been offered that the pressure in the inner ear brought about by the response of the ear to high level sound is sufficiently great to destroy portions of the delicate tissues. Others have advanced the theory that heat transfer within the end organ is destructive to the sensory cells.

An additional mechanism which must be considered is that of anoxia caused by the reduction in blood supply secondary to capillary constriction. Although not necessarily deemed to be the only mechanism of cochlear damage, anoxic conditions in the end organ regions probably account for some damage. It remains to be proven whether this recently noticed phenomenon is one of the major determinants of cochlear and vestibular malfunction in the wake of noise stimulation.

Other Indicators

There exist several additional realms in which one might list various indications of environmental noise. The following areas are much more difficult to quantify than the topics presented thus far, for relatively little information has been provided in each. The topics are quite subjective in nature, and they greatly overlap.

Educational indicators of environmental noise. Various physiological alterations can be considered to have a deleterious effect upon motor performance and productivity. Research evidence in this realm, however, is not conclusive, causing Kryter[65] to state that noise is sometimes reported to have a good effect, a bad effect, or no effect.

In tasks that demand great concentration and rapid reaction, noise interference has been demonstrated. Through studies of vigilance, it was noted that subjects who were monitoring complicated racks of dials and gauges were less vigilant when noise was introduced into the environment.[58] Broadbent[21] suggested that noise causes brief lapses in attention or creates a situation of cortical over-arousal resulting in reduced behavioral control. These observations have been taken to indicate that efficiency could be reduced by high noise levels.

The effects of noise upon pupil performance has been a subject of some study with, understandably, differing conclusions. On the one hand, some say that noise does not hinder pupil performance,[111] whereas others have arrived at the conclusion that pupil performance is negatively influenced.[134]

Of particular note is the fact that several public schools have been closed in recent years as a direct result of aircraft noise originating from major air terminals.[119] Perhaps this is sufficient evidence that high level noise, which seems to be encroaching upon the educational units in the community, has a deleterious effect upon some aspects of performance and the learning process.[22]

Certain noises by their intensity, interest factor, or rhythm will serve as interfering agents and should be considered potentially disruptive to normal educational situations.[100]

Safety as an indicator of environmental noise. The various physiological upheavals which are laid to noise may work singly or in combination to decrease an individual's vigilance creating an accident situation.[59,83] For instance, pupillary dilation in response to noise exposure creates an artifactual change in color perception.[57] It is speculated that these chromatic inconsistencies may create a high accident risk in those activities in which accurate visual perception is essential.

Untoward reactions in the vestibular system during intense sound stimulation may create in the individual an unsteadiness sufficient to place him in jeopardy — especially if he is a steeplejack.

With little imagination, one could list great numbers of ways in which a noisy environment can pose a threat to the health and safety of the persons therein. The limitations of time and space do not permit further elucidation with the exception of a very interesting paradox which is becoming recognized.

The American automobile industry has been striving during the past few years to develop a product which provides a smoother *and quieter* ride. With the effective use of undercoating, special suspension systems, carpeting, acoustic treatment of the top liner, padding, and isolation of especially noisy areas, the modern automobile is considerably less noisy to the occupant. The acoustic barrier afforded the rider insulates him from disturbing and distracting outside noises — including warning signals.

A car traveling at a moderate speed, with its windows rolled up and the radio and air conditioner operating, may develop an ambient noise level of approximately 90 dB. The shell of the car provides at least 20 dB of attenuation (sound reduction); thus, a warning signal such as a siren or train whistle must impinge upon the outside of the car at a sound level of at least 110 dB for the driver to be barely able to hear the sound. Of course, the warning signal must be heard well in order to serve its function; so it should strike the car at levels between 120 and 130 dB. A popular siren for police cars is rated to produce sound at a distance of 100 ft measuring 100 dB. In the above acoustical situation, there is no way the driver of an automobile can be auditorily aware of the onrushing emergency vehicle.

A recent court case was completed in which the survivors of a man killed at a railroad crossing received an award of $85,000. A strong point made by the plaintiff was that there was little evidence that the deceased driver heard the approaching train.

The foregoing illustration indicates that the abatement of noise is sometimes a two-edged sword. On the one hand, the safety factor of automobiles is improved by reducing the fatiguing noise of the car in motion, but paradoxically, the improvements place the driver and other occupants of the car in jeopardy with respect to hearing warning signals.

Psychological indicators of environmental noise. Any stimulus which has an effect upon physiological balances is bound to be psychologically impressive as well. Little is known, however, about the emotional concomitants of noise exposure. Allusions are frequently made concerning the frustrations, fatigue, vexations, irritations, and immobilizing qualities of high intensity sounds; but most of these observations are subjectively based.[27] Westman[132] has discussed in some detail the role of noise as a disruptive influence in the home. He observes that "togetherness at the supper table is hampered by household noises and by the general tenseness fanned by the daylong din." He continues by relating that "we don't understand that noise makes us less efficient, less effective and more tense. Instead, we scapegoat. We take our tensions out on each other. Mothers yell at the youngsters. Parents bicker."

There would probably be little argument that mental state is greatly influenced by physical well-being. Therefore, if the internal functions of the body have been tossed into a form of disarray as a reaction to sound stimulation, the person's psychological condition can be expected to undergo some modification, which may or may not be overt. Psychological reactions can be related to the noxious aspect of the sound source, to the relative pleasure— displeasure an individual is experiencing at the onset of the noise, and to the basic anxiety level of the person or his evaluation of his total situation at the time noise occurs.[13]

Anger, a very overt personal reaction to an irritant, is becoming more commonly expressed toward various community noises. Often, anger begets action — thus, class suits are appearing on legal dockets in which one person or a group of persons are suing the source of community noise. For example, over $10 billion in such suits have been filed against the Los Angeles International Airport alone.[119]

The psychophysical techniques already discussed are predicated upon the reality of individual variation with respect to stimulus response. The sensation created by a given stimulus, regardless of the sensory modalities, is modified by the state of the whole being. Very irritable persons will, therefore, be more irritated by a particular type of noise, as a group, than will low-key individuals.[18] Persons engaged in very precise activities will find noisy conditions more distracting than will laborers involved in gross tasks.[58]

It is readily apparent that these variations pose a severe limitation to the predictive success of attempts to scale the psychological responses to given sounds.

Social indicators of environmental noise. Glorig[47] coined the term "socio-cusis" to describe the condition of an individual who has sustained a permanent hearing loss as a result of non-occupational noise exposure. It connotes that there are social implications of community noise. Questions which might be asked in this realm include:

a. What is the role of noise in the environment as one of the dehumanizing influences in inner city life?

b. Is there a relationship between noise levels and the apparent social upheavals taking place in heavily populated urban areas?

c. A large proportion of young people who were activists in the recent peace movement were from large cities. To what extent did the high ambient noise levels in their environment create in them a drive to change their world?

d. How does the noise factor relate to other features in the crowding problems of over-populated areas?

e. If noise abatement were effectively initiated across the country, what possible effects would there be on the social climate of the population?

f. What are the leading noise irritants in various areas of the country? In 1956, a newspaper survey in New York City indicated that the most unsavory of all noise sources was the clanging garbage can.[114] Other surveys have listed truck noise, aircraft noise, traffic noise,[26] the sounds of construction equipment, lawnmowers, trains and subways, and appliances.[9]

The social implications of environmental noise have been subjected to insufficient concentrated study. Hopefully, the above questions and many more will be the target of scientific investigation during the upcoming decade.

Political indicators of environmental noise. The population of most industrialized nations is beginning to realize that noise-producing equipment and gadgetry are undesirable elements in their life.[54] That this is true can be attested to on a number of political fronts.[30] The public is demanding that the continuing spiral of noise levels be not only stopped but rolled back. To this end, they are becoming less tolerant of technologic developments which are predicted to raise the overall noise level.[68] A classic example of this factor is the controversy over the supersonic transport project (SST). In speaking against the SST continuation funds, Senator Proxmire stated:[32]

"... the SST is the clearest example of private power and capital versus the public interest that I have seen in my 13 years in the U. S. Senate. No one wants or needs the SST except big labor, big industry, and big politics. The public interest is the decided underdog. Whether it can prevail against such overwhelming odds remains to be seen."

The SST project was abandoned along with the jobs of several thousand aerospace workers. A major reason for such a response of the elected representatives of government was that they had received a large number of letters and telegrams from their constituents insisting that they stop a project that might be projected to produce another noisy device.

Other political indicators of environmental noise can be seen in the increasing interest of the govenmental regulating agencies in curbing the noise emitted by various objects. In December of 1970, the Federal Trade Commission listed noisy toys in their group of playthings that should receive serious review prior to their widespread manufacture and distribution.

The Walsh-Healy Revision of 1969 (Ref. 42) and, later, the Williams-Steiger Occupational Safety and Health Act of 1971 (Ref. 43) have defined the allowable noise exposure for employees in nearly every type of occupational pursuit.

The Environmental Protection Agency has held hearings on numerous aspects of the noise question in order to determine its role in the control of appliances and other products with respect to the noise output of the devices.

Few other areas of public concern have such overwhelming political support as that presently enjoyed by anti-noise interest. As witness the SST rejection and other similar dramatic reversals in national policy, there are certain political indicators of environmental noise that are difficult to quantify but which, nonetheless, are quite evident in government at all levels.[94]

Economic indicators of environmental noise. It has been estimated that American industry loses $4 billion annually as a direct result of noise through noise-related accidents, turnover, and absenteeism.[119] Such a jolt to the economy can hardly be overlooked. In addition, there are other economic indicators which might be suggested.

A few years ago, a vacuum cleaner was marketed which was touted to be a "quiet" appliance. Unfortunately, it was ahead of its time, for it failed to receive consumer acceptance. Housewives couldn't believe that it was sweeping effectively if it made so little noise.

Also, a typewriter was developed which had a special roller designed to absorb the clatter of the keys against the paper. It too was rejected. Secretaries were afraid to lose the tell-tale clangor of their typing machines, feeling their bosses would accuse them of not working with their former efficiency.

A common thread through both of the above anecdotes is the apparent equation of noise with power and effectiveness. A motorcyclist chooses to strip his machine of the muffler to sense the power of the motor vicariously through his hearing sense. The hot rodder does the same. Musicians have developed amplification systems that give a heretofore unrealized acoustic power output.

The trend appears to be that the economic cutting edge has been in the direction of sound producing equipment with little regard on the part of the consumer for paying the necessary higher prices for items with noise emission reduced.[14]

The political directions taken in the late 1960's and early 1970's may reverse the trend so that it will become more expensive to produce and market noisy items because of taxes or penalties against products which do not meet noise specifications.

CONCLUSION

Noise is a common problem whose impact cuts across nearly every facet of human existence. The complexity of the problem defies circumscription into a small number of universal indicators, at least at the present status of knowledge of noise and its effects.

The human indicators come largely after the facts. There is a temporary delay in the physiological effect that may accumulate over years, resulting in one or more physical problems. The loss of hearing, likewise, does not occur immediately upon exposure to high level sound. It, too, becomes more conspicuous as the bits of structural damage accumulate to cause a severe communication problem in later life.

Whether noise and its many effects can be reduced to a single element indicator is yet to be thoroughly tested. At present, there are a large number of diverse elements which appear to defy such an approach.

SUMMARY

A number of potential indicative features of noise in the environment have been discussed in an effort to illuminate the factors that give the most clear definition of noise response. A brief summarizing statement of each element discussed may be given as follows:

1. The noise problem is yet to be sufficiently defined in order to understand its depth and breadth.
2. Physical indicators are quite reliable and the technology is currently sufficient to provide highly refined sound analysis.
3. A combination of physical and psychophysical indicators of noise in the environment are inadequately determined so far, in that single-unit descriptions of the perceptual characteristics of noise do not yet have universal application.
4. Noise is a stressor in the classic physiologic sense.
5. Intense sound has a damaging effect upon the sensory cells of the auditory region.
6. Insight into the hurtful quality of high level sound is shown in the reduction of blood supply to the auditory and vestibular end organs, providing one possible explanation of the mechanism of noise destruction of ear tissues as being anoxia.
7. Such areas as education, safety, psychology, social sciences, politics, and economics offer further indications of noise in the environment but their further clarification is essential.

REFERENCES

1. American Academy of Opthalmology and Otolaryngology, Guide for conservation of hearing in noise (revised), 1969.

2. American Industrial Hygiene Association, *Industrial Noise Manual*, 1966.
3. American National Standards Institute, Physical measurement of sound, method for. S1.2–1962.
4. American National Standards Institute, General-purpose sound level meters, specifications for. S1.4–1961.
5. American National Standards Institute, Preferred frequencies and band numbers for acoustical measurements. S1.6–1967.
6. American National Standards Institute, Octave, half-octave, and third-octave band filter sets, specifications for S1.11–1966.
7. American National Standards Institute, Specifications for audiometers, S3.6–1969.
8. Anson, B. and C. B. McVay. *Surgical Anatomy.* W. B. Saunders Co., Philadelphia, 1971.
9. Apps, D. Cars, trucks and tractors as noise sources, in *Noise as a Public Health Hazard – Proceedings of the Conference,* W. D. Ward and J. E. Fricke (eds.). The American Speech and Hearing Association, Washington, D. C., 1968.
10. Arguelles, A. E., D. Ibeas, J. Pomes Ottone, and M. Chekherdemian. Pituitary-adrenal stimulation by sound of different frequencies. *J. Clin. Endocrin.* 22: 846–852 (1962).
11. Arguelles, A. E., M. A. Martinez, E. Pucciarelli, and M. V. Disisto. Endocrine and metabolic effects of noise in normal, hypertensive and psychotic subjects, in *Physiological Effects of Noise,* B. Welch and A. Welch (eds.). Plenum Press, New York, 1970.
12. Atherley, G. R. C., S. L. Gibbons, and J. A. Powell. Moderate acoustic stimuli; the interrelation of subjective importance and certain physiological changes, *Ergonomics,* 13 (5): 536–545 (1970).
13. Azrin, N. H. Some effects of noise on human behavior, *J. Exp. Anal. Behav.,* 1: 183–200 (1958).
14. Baron, R. A. *The Tyranny of Noise.* St. Martin's Press, New York, 1970.
15. Bergman, M. Hearing in the Mabaans. A critical review of related literature, *Arch. Otol.,* 84: 411–415 (1966).
16. Baranek, L. L. *Noise Reduction.* McGraw-Hill Book Company, New York, 1960.
17. Bishop, D. E. and R. D. Horonjeff. Procedures for developing noise exposure forecast areas for aircraft flight operations (Rept. DS-67-10, Contract FA 67WA-1705). Bolt, Beranek, and Newman, Inc., Van Nuys, California, 1967.
18. Borsky, P. N. Effects of noise on community behavior, in *Noise as a Public Health Hazard – Proceedings of the Conference*, W. D. Ward and J. E. Fricke (eds.). The American Speech and Hearing Association, Washington, D. C., 1968.
19. Bredberg, G. Cellular pattern and nerve supply of the human organ of Corti, *Acta. Oto. Laryngol.,* Supplement 236 (1968).

20. Bredberg, G., H. H. Lindeman, H. W. Ades, R. West, and H. Engstrom. Scanning electron microscopy of the organ of Corti, *Science* 170: 861–863 (1970).

21. Broadbent, D. E. Effects of noise on behavior, in *Handbook of Noise Control,* C. M. Harris (ed.). McGraw-Hill, New York, 1957.

22. Brody, J. F. Jr. Conditioned suppression maintained by loud noise instead of shock, *Psychon. Sci.* 6(1): 27–28 (1966).

23. Buckley, J. P., and H. H. Smookler. Cardiovascular and biochemical effects of chronic intermittent neurogenic stimulation, in Physiological Effects of *Noise,* B. Welch and A. Welch (eds.). Plenum Press, New York, 1970.

24. Burns, W. *Noise and Man.* Lippincott Publishers, Philadelphia, 1968.

25. California, State of. Noise control safety orders, Div. Indust. Safety, Dept. Indust. Relations, 1962.

26. Chalupnik, J. D. (ed.) *Transportation Noises.* Univ. of Washington Press, Seattle, 1970.

27. Cohen, A. Effects of noise on psychological state, in *Noise as a Public Health Hazard – Proceedings of the Conference,* W. D. Ward and J. E. Fricke (eds.). American Speech and Hearing Association, Washington, D. C., 1968.

28. Cohen, A., J. Anticaglia, and H. H. Jones. "Sociocusis"–hearing loss from non-occupational noise exposure, *Sound Vib.*, 4 (11): 12–20 (1970).

29. Coles, R. R. A., et al. Hazardous exposure to impulse noise, *J. Acous. Soc. Am.* 43: 336–343 (1968).

30. Commerce, Department of. *The Noise Around Us: Including Technical Backup,* Report CTAB-71-1 (1970).

31. Compton, R. L. An investigation of the oto-hazardous potential of high school band music, unpublished Master's Thesis, The University of Tennessee, Knoxville, Tennessee.

32. Congressional Record, 117 (31): 52636 (March 9, 1971).

33. Crum, M. A. The incidence of hearing loss among 1000 ninth-grade students as related to environmental noise exposure, unpublished Master's Thesis, The University of Tennessee, Knoxville, 1968.

34. Darner, C. L. Sound pulses and the heart, *J. Acoust. Soc. Am.*, 39 (2): 414–416 (1966).

35. Davis, H., and S. R. Silverman. *Hearing and Deafness.* Holt, Rinehart and Winston, New York, 1970.

36. Davis, R. C. and T. Berry. Gastrointestinal reactions to response-contingent stimulation, *Psychol. Rep.*, 15: 95–113 (1964).

37. Davis R. B., A. M. Buchwald, and R. W. Frankman. Autonomic and muscular responses and their relation to simple stimuli, *Psychol. Monographs,* (1955).

38. Dey, F. L. Auditory fatigue and predicted permanent hearing defects from rock-and-roll music, *New Eng. J. Med.*, 282 (9): 467–470 (1970).

39. Dickson, E. D. D., and D. L. Chadwick. Observations on disturbances of

equilibrium and other symptoms induced by jet engine noise, *J. Laryngol. Otol.*, 65: 154–165 (1951).

40. Downs, M. P., W. G. Hemenway, and M. E. Doster. Sensory overload, *Hearing and Sp. News,* May 1, 1969, p.11.

41. Engström, H., H. W. Ades, and A. Andersson. *Structural Pattern of the Organ of Corti.* The Williams & Wilkins Co., Baltimore, 1966.

42. *Federal Register*, 34 (96): 7948–7949 (May 20, 1969).

43. *Federal Register*, 36 (105): 10518 (May 29, 1971).

44. Fletcher, H., and W. A. Munson. Loudness, its definition, measurement and calculation, *J. Acoust. Soc. Am.* 24: 80 (1933).

45. Flugrath, J. M. Modern-day rock-and-roll music and damage-risk criteria, *J. Acoust. Soc. Am.*, 45 (3): 704–711 (1969).

46. Glorig, A. *Noise and Your Ear.* Grune & Stratton, New York, 1958.

47. Glorig, A. Guidelines for noise exposure control, *Am. Ind. Hygiene Assn. J.*, 28: 418–424 (1967).

48. Glorig, A., W. D. Ward, and J. Nixon. Damage risk criteria and noise-induced hearing loss, *Arch. Oto.*, 74: 413–424 (1961).

49. Goldstein, S. N. A prototype standard and index for environmental noise quality, presented to the Acoustical Society of America, October 21, 1971; Report MTP-358, The Mitre Corporation.

50. Hale, H. B. Adrenalcortical activity associated with exposure to low frequency sounds, *Am. J. Physiol.*, 171: 732 (1952).

51. Harris, C. (ed.). *Handbook of Noise Control.* McGraw-Hill Book Company, New York, 1957.

52. Health, Education and Welfare, Department of. *Hearing Levels of Children by Age and Sex.* National Center for Health Statistics, Series 11, Number 102, 1970.

53. Henry, K. R., and R. E. Bowman. Early exposure to intense acoustic stimuli and susceptibility to audiogenic seizures, in *Physiological Effects of Noise.* B. Welch and A. Welch (eds.). Plenum Press, New York, 1970.

54. Horning, D. F. *Noise – Sound Without Value.* Committee on Environmental Quality, Federal Council for Science and Technology, 1968.

55. Hosey, A. D. and C. Powell (eds.). *Industrial Noise – A Guide to its Evaluation and Control*, PHS Pub. No. 1572, 1967.

56. International Organization for Standardization, Standard reference zero for the calibration of pure tone audiometers, ISO recommendation R 389, 1964.

57. Jansen, G. Effects of noise on physiological state, in *Noise as a Public Health Hazard – Proceedings of the Conference,* W. D. Ward and J. E. Fricke (eds.). The American Speech and Hearing Association, Washington, D. C., 1968.

58. Jerison, H. J. and S. Wing. Effect of noise and fatigue on a complex vigilance task, WADC Tech. Rep. TR-57-14, Wright Patterson AFB, 1967.

59. Jones, H. H. and A. Cohen. Noise as a health hazard at work, in the community, and in the home, *Pub. Health Rep.*, 83 (7): 533–536 (1968).

60. Knudsen, V. O. Acoustics in comfort and safety, *J. Acoust. Soc. Am.*, 21 (4): 296–301 (July 1949).

61. Knudsen, V. O. Noise, the bane of hearing, *Noise Control,* 1 (3): 11–13 (May 1955).

62. Kohonen, A. Effect of some ototoxic drugs upon the pattern and innervation of cochlear sensory cells in the guinea pig, *Acta-Oto-Laryng.* Suppl. 208 (1965).

63. Kryter, K. D. Scaling human reactions to the sound of aircraft, *J. Acoust. Soc. Am.*, 35: 866 (1963).

64. Kryter, K. D. Modifications of noy table, *J. Acoust. Soc. Am.,* 36: 394 (1964).

65. Kryter, K. *The Effects of Noise on Man.* Academic Press, New York, 1970.

66. Kryter, K. D. and K. S. Pearsons. Some effects of spectral content and duration of perceived noise level, *J. Acoust. Soc. Am.*, 35: 866 (1963).

67. Kryter, K., et al. Hazardous exposure to intermittent and steady-state noise, *J. Acoust. Soc. Am.,* 39: 451–464 (1966).

68. Lang, J. and G. Jansen. *The Environmental Health Aspects of Noise Research and Noise Control.* World Health Organization EURO 2631, Copenhagen, 1970.

69. Lawrence, M. Effects of interference with terminal blood supply on organ of Corti, *Laryngoscope*, 76 (8): 1318–1337 (1966).

70. Lawrence, M. Circulation in the capillaries of the basilar membrane, *Laryngscope*, 80 (9): 1364–1375 (1970).

71. Lawrence, M., G. Gonzalez, and J. E. Hawkins, Jr. Some physiological factors in noise-induced hearing loss, *Am. Indust. Hyg. Assn. J.*, 28 (1967).

72. Lebo, C. P. and K. Oliphant. Music as a source of acoustic trauma, *Laryngoscope*, 78 (7): 1211–1218 (1968).

73. Lebo, C. P. and K. Oliphant. Music as a source of acoustic trauma, *J. Audio. Eng. Soc.*, 17 (5): 535–538 (1969).

74. Levy, L. Sympatho-adrenomedullary responses to emotional stimuli: methodologic, physiologic and pathologic considerations, in *An Introduction to Clinic Neuroendocrinology.* E. Bajusz (ed.). S. Karger, Basel, 1967.

75. Lim, D. and W. Melnick. Acoustic damage of the cochlea, *Arch. Otol.,* 94: 294–305 (1971).

76. Lindemann, H. Results of speech intelligibility survey in cases of noise traumata, *Int. Audiol.*, 8 (4): 626–632 (1969).

77. Lipscomb, D. M. Cochlear damage resulting from exposure to high intensity "hard rock" music, presented at a meeting of the Am. Acoust. Soc., Philadelphia, 1969.

78. Lipscomb, D. M. Ear damage from exposure to rock and roll music, *Arch. Otol.*, 90: 545–555 (1969).

79. Lipscomb, D. M. High intensity sounds in the recreational environment: a hazard to young ears, *Clin. Pediat.*, 8 (2): 63–68 (1969).

80. Lipscomb, D. M. Noise in the environment: The problem, *Maico Audiological Library Series*, 8 (1), 1969.

81. Lipscomb, D. M. Noise in the environment: experimental results, *Maico Audiological Library Series*, 8 (2), 1969.

82. Lipscomb, D. M. Increase in cochlear cell damage as a function of variations in exposure to "acid rock" music, presented at a meeting of the Am. Acoust. Soc., Atlantic City, 1970.

83. Lipscomb, D. M. Important considerations in the use of hearing conservation guidelines, *Maico Audiological Library Series*, 8 (5), 1970.

84. Lipscomb, D. M. Non-occupational noise: a growing problem, *Natl. Hearing Aid J.*, (January 1970).

85. Lipscomb, D. M. The noise problem 1970: A status report, presented to the State of Tennessee Legislative Council Committee, April 1970.

86. Lipscomb, D. M. The increase in prevalence of high frequency hearing impairment among college students, *Audiology*, 11: 231–237 (1972).

87. Lipscomb, D. M. Noise as a source of technogenic disease, presented to the National Conference on Technogenic Diseases Buffalo, N. Y., April 24, 1971.

88. Lipscomb, D. M. Non-occupational noise and the effect upon hearing of young persons, pp. 290–316 in *Noise Control*. Hearings before the Subcommittee on Public Health and Environment of the Committee on Interstate and Foreign Commerce, House of Representatives. Serial No. 92–30, 1971.

89. Lipscomb, D. M. Physiological and psychological effects of aircraft noise, presented to Short Course on Aircraft Noise, Aachen, W. Germany, April 2, 1971.

90. Lipscomb, D. M. Considerations in perceived 'noisiness,' *J. Tenn. Sp. & Hrg. Assn.* 16 (2): 28–31 (1972).

91. Lipscomb, D. M. and R. L. Willhoit. Differential destruction of cochlear cells by low, medium and high frequency octave bands of noise, presented at a meeting of the Am. Speech and Hearing Assn., November, 1970.

92. Litke, R. E. Elevated high-frequency hearing in school children, *Arch. Otol.*, 94: 255–257 (1971).

93. Majeau-Chargois, D. A., C. I. Berlin, and G. D. Whitehouse. Sonic boom effects on the organ of Corti, *Laryngoscope*, 80 (4): 620–630 (1970).

94. Mayor's Task Force on Noise Control, *Toward a Quieter City*. Board of Trade, New York, 1970.

95. Navarra, J. G. *Our Noisy World*. Doubleday & Co., Garden City, 1969.

96. Owen, C. The incidence of hearing loss as related to environmental noise exposure among 1000 sixth-grade students, unpublished Master's thesis, The University of Tennessee, Knoxville, 1968.

97. Pell, S. and T. H. Dickerson. Changes in hearing acuity of noise-exposed women, *Arch. Otol.*, 83: 207–212 (1966).

98. Peterson, A. P. G., and E. E. Gross. *Handbook of Noise Measurement.* General Radio Company, West Concord, Mass., 1967.

99. Poche, L. B., C. W. Stockwell, and H. W. Ades. Cochlear hair-cell damage in guinea pigs after exposure to impulse noise, *J. Acoust. Soc. Am.,* 46 (4): 947–951 (1969).

100. Rettinger, M. *Acoustics – Room Design and Noise Control.* Chemical Publishing Co., Inc., New York, 1968.

101. Rintelmann, W. F. and J. F. Borus. Noise-induced hearing loss and rock and roll music, *Arch. Otol.,* 88: 377–385 (1968).

102. Robinson, D. W. Towards a unified system of noise assessment, *J. Sound Vib.,* 14: 279–298 (1971).

103. Rosen, S., M. Bergman, D. Plestor, A. ElMofty, and M. Satti. Presbycusis study of a relatively noise-free population in the Sudan, *Ann. Otol. Rhinol. Laryngol.,* 71: 727–743 (1962).

104. Rosen, S. and J. Olin. Hearing loss and coronary heart disease, *Arch. Otol.,* 82: 236–243 (1965).

105. Rosen, S., D. Plestor, A. ElMofty, and H. V. Rosen. High frequency audiometry in presbycusis: a comparative study of the Mabaan Tribe in the Sudan with urban populations, *Arch. Otol.,* 79: 18–32 (1964).

106. van der Sandt, W., A. Glorig, and R. Dickson. A survey of the acuity of hearing in the Kalahari bushman, *Int. Audiol.,* 8 (2): 290–298 (1969).

107. Schmidt, P. H. Presbyacusis and noise, *Int. Audiol.,* 8 (2): 278–280 (1969).

108. Selye, H. The general adaptation syndrome and the diseases of adaptation, *J. Clin. Endocrin.,* 6: 117–230 (1946).

109. Selye, H. Stress and disease, *Science,* 122: 625–631 (1955).

110. Selye, H. *The Stress of Life.* McGraw-Hill, New York, 1956.

111. Slater, B. R. Effects of noise on pupil performance, *J. Educ. Psychol.,* 59: 239–243 (1968).

112. Smitley, E. K. and W. F. Rintelmann. Continuous versus intermittent exposure to rock and roll music, *Arch. Environ. Health,* 22: 413–420 (1971).

113. Sontag, L. W. Effect of noise during pregnancy upon fetal and subsequent adult behavior, in *Physiological Effects of Noise.* B. Welch and A. Welch (eds.). Plenum Press, New York, 1970.

114. Soroka, W. W. Community noise surveys, in *Noise as a Public Health Hazard – Proceedings of the Conference.* W. D. Ward and J. E. Fricke (eds.). The American Speech and Hearing Assn., Washington, D. C., 1968.

115. Stern, R. M. Effects of variation in visual and auditory stimulation on gastrointestinal motility, *Psychol. Rep.,* 14: 799–802 (1964).

116. Stevens, S. S. (ed.). *Handbook of Experimental Psychology.* Wiley Publications, New York, 1951.

117. Stevens, S. S. The measurement of loudness, *J. Acoust. Soc. Am.,* 27: 815–829 (1955).

118. Stevens, S. S. and H. Davis. *Hearing.* Wiley Publishers, New York, 1938.

119. Still, H. *In Quest of Quiet.* Stackpole Publishers, Harrisburg, 1970.
120. Stockwell, C. W., H. W. Ades, and H. Engström. Patterns of hair cell damage after intense auditory stimulation, *Ann. Otol. Rhino. Laryngol.,* 78 (6): 1144–1170 (1969).
121. Taylor, G. D., and E. Williams. Acoustic trauma in the sports hunter, *Laryngoscope,* 76: 863–879 (1966).
122. Thiessen, G. J. Effect of noise from passing trucks on sleep, presented to the Acoustical Society of America, April, 1969.
123. Trent, H. M. and B. Anderson. *Glossary of Terms Frequently used in Acoustics.* Am. Institute of Physics, New York, 1960.
124. U. S. Air Force Medical Service. Hazardous noise exposure, AF Reg. no. 160-3, Washington, D. C., 1956.
125. Ward, W. D. Damage risk criteria for line spectra, *J. Acoust. Soc. Am.,* 34: 1610–1619 (1962).
126. Ward, W. D. Auditory fatigue and masking, Chap. 7 in *Modern Developments in Audiology.* J. Jerger (ed.). Academic Press, New York, 1963.
127. Ward, W. D. Adaptation and fatigue, Chap. 9 in *Sensorineural Hearing Processes and Disorders.* A. B. Graham (ed.). Little, Brown Publishers, Boston, 1967.
128. Ward, W. D. Susceptibility to auditory fatigue, in *Advances in Sensory Physiology,* Vol. 3. W. D. Neff (ed.). Academic Press, New York, 1967.
129. Ward, W. D., and J. E. Fricke (eds.). *Noise as a Public Health Hazard – Proceedings of the Conference,* The American Speech and Hearing Association, Washington, D. C., 1968.
130. Ward, W. D., A. Glorig, and D. L. Sklar. Temporary threshold shift from octave-band noise: Applications to damage-risk criteria, *J. Acoustic Soc. Am.,* 31: 522–528 (1959).
131. Welch, B. and A. Welch (eds.). *Physiological Effects of Noise.* Plenum Press, New York, 1970.
132. Westman, J. C. Interview with Bruce Ingersoll, Chicago Sun-Times Service, 1971.
133. Wood, W. S. An investigation of the continuous sound levels available from stereo components, unpublished Master's Thesis, The University of Tennessee, Knoxville, Tennessee.
134. Wrightson, P. A review of research on noise, with particular reference to schools, Quarterly Bulletin No. 8, pp. 20–27, GLC Research and Intelligence Unit (London), 1969.
135. Yeiser, J. L. An incidence survey of hearing impairments among 1000 high school seniors as related to environmental noise exposure, unpublished Master's Thesis, The University of Tennessee, Knoxville, 1969.
136. Zwicker, E., G. Flottorp, and S. S. Stevens. The critical bandwidth in loudness summation, *J. Acoust. Soc. Am.,* 29: 548–557 (1957).

DEVELOPING A SOIL QUALITY INDEX

Richard H. Rust, Russell S. Adams, Jr. and William P. Martin

Professors & Head, Department of Soil Science
University of Minnesota
St. Paul, Minnesota 55101

The possibility exists that a kind of "soil quality index" might be developed which would relate to the environmental impact of continuing or sustained use of chemical amendments in crop production or certain other land uses. Present concerns are mostly related to use of soluble fertilizers, including nitrogen and phosphorus; applications of weedicides and pesticides; and distribution of industrial and domestic wastes that may also include heavy metals in significant concentrations.

For purposes of illustrating a working concept, let us consider development of a "soil quality index" relevant to use of soluble nitrogen forms. First, we must detail the probable factors that need to be measured in the soil and root-growing environment. One can visualize a kind of input–output system where *inputs* will be contributions principally from crop residues, mineralization of soil organic matter, animal manures, and chemical amendments and *outputs* are in the form of crop removal (seed and stalk), leaching losses, ammonification, and gaseous losses. Additionally, a "holding" or "residence" time in soil for N must be evaluated in terms of exchange capacity or absorptive capacity of the soil to some depth. This effective depth may be related to water–table position and character.

Towards the ideal condition, the objective would be to establish a "zero balance" in which N utilized or retained in the soil would be equal to the amount added or available through mineralization.

Different crops have different N requirements. Moreover, the same crop will have different N requirement depending on the associated nutrient availabilities. Finally, the same crop will have changing seasonal demand for N depending on stage of growth and rate of growth as it might be generally related to photosynthetic processes.

Soils also differ in respect to their retention of N as a result of a wide range in textures (and, therefore, absorptive capacities) and as a result of temperature, moisture, and aeration regimes. In a specific soil, these parameters can be indicated with some degree of quantification.

The retention of N in any soil is also related to the form of N, e.g., ammonium ion, nitrate ion, or organic complex. The form of N will generally be more difficult to specify.

The soil must also be related to its climatic environment. The amount and distribution of precipitation, the march of soil temperature, and often the microclimatic factors of aspect and slope must be considered.

To place the problem of developing the soil quality index in some quantitative reference, we may, by analogy, consider the concepts inculcated in the Smith and Wischmeier[4] equation that attempts to predict soil loss, on an average, for a specified rotation. The equation form is

$$A = R\ K\ LSCP$$

where the respective letters relate to functional relationships between A, soil loss (in the growing season); R, rainfall; K, soil erodibility; L, length of slope; S, steepness; C, cropping system; and P, erosion control practice, respectively.

Since N loss or excess will largely be removed by water (although there may be instances of significant gaseous losses), we might develop the analogous loss figure in terms of pounds of N (or NO_3) for a specified land use under specified management.

The rainfall factor that must be evaluated in the soil loss equation is also a consideration in nitrogen losses. Probability statements regarding average expectation of growing season precipitation in a given locale should be developed. Further elaboration of distribution and intensity would be helpful.

Whereas soil erodibility, K, is a significant parameter (i.e., the ease of particle detachment) in the soil loss equation and is an intrinsic property of the particular soil, a more appropriate parameter for N, and also intrinsic, would be permeability of the rooting zone to water. While the general concept may include both water and air permeability, it is probable that water permeability would be the more useful value. However, considerable difficulty will arise in distinguishing water permeability in the saturated versus the unsaturated soil and relating it to the movement of soluble N.

While length of slope, L, and steepness, S, are important criteria in the soil loss relationship, they would probably be less significant in the N loss analysis except as losses would relate to surface runoff. It would seem reasonable to initially assume that N compounds are, or would be, incorporated in the surface horizon.

Position and nature of water table in respect to landscape would be a relevant factor in N movement. Nearness to rooting zone, the static or dynamic nature of the water table, and ionic composition of the water — all would bear on the N loss or "capture" in a selected situation.

The cropping system, C, assumes great importance in N evaluation. N consumption will be a function of type of crop and quality of growth. Likewise, in the case of legumes, N production must be considered. Complete or partial crop removal and nature of residue management would need to be considered.

Erosion control practices, P, while more directly related to soil loss, must also be considered as effective in water control. Thus, if a degree of erosion control results in a reduction of surface water runoff, there is a consequent increase in infiltrating water which, in turn, will effectively move soluble N forms downward or laterally in the soil.

In addition to the kinds of consideration used in the soil loss equation, some emphasis must be given to microbiologic activity as a measure of mineralization or immobilization in a particular soil. While this may not always be predictable, it is functionally related to soil temperature and soil moisture regimes which can be derived in some predictable manner.

The foregoing may seem more closely related to an attempt to predict an N excess or deficit in a given site and not so directly related to a "soil quality index". However, it is readily apparent that, as in the Wischmeier equation, certain soil properties will need to be measured and evaluated. And some range in these intrinsic, and also extrinsic, properties will establish a kind of "soil quality index." The quality index, along with cropping and climatic factors, can then be used for decision making.

It does, however, seem apparent that a kind of landscape, or perhaps watershed, basis of evaluation ultimately is needed. We are attempting to derive an estimate of, say, N addition to a given water body or stream, and it will be the sum total of additions of groundwater flow originating from individual soils in the landscape and the cropping and fertility management practices being used.

To establish an N soil quality index in a specific environment, the research will necessarily have to be of some factorial design in view of the wide range of soil, crop, climatic, and management conditions. A closed, or limited, watershed where either control of the variables or attainable dynamic measurements thereof are possible would seem to be necessary. While it would generally be impossible to impose uniform cropping or other management procedures on a watershed basis, unless very small, the nearer this condition could be established the more predictive the results.

Research sites should be established in a number of areas as an attempt to represent the natural variability of soils and of climate. Labelled N might be used in some situations to distinguish N sources and establish their relationships to ultimate N concentration in the monitored water.

A specific contribution of soil science in this analysis would be to establish the functional relationship of N transmission in the rooting zone of individual soils of the landscape. By using a taxonomic consideration in the delineation of the watershed (or landscape), certain properties of permeability, microbiologic activity, ion composition, and absorptive capacity become stratified in the factorial analysis.

The research must be conducted over time, not only to sample climatic variability but also to establish rates of accumulation or depletion and, as was mentioned earlier, "residence" time of certain N compounds.

An example of one possible index will be illustrative. In terms of pesticide residues, the factors previously discussed (soil variation, climatic variation, and sediment loss) also apply. Because of their extremely low solubility, the movement of pesticides from the land into surface bodies of water is primarily a function of sediment loss. Therefore, the Wischmeier equation should be readily adaptable to an estimation of pesticide pollution of surface waters from agricultural practices.

Estimation of a "soil quality index" becomes more complex. The expression of pesticide residue injury to a plant or the occurrence of unwanted residues in a plant is a function of the chemical properties of a given pesticide; climatic conditions, including rainfall and soil temperature; the strength of sorption of the pesticide by the soil; and the ability of the crop to degrade or translocate a given pesticide. Thus, crop species and variety also become a factor.

A simple and gross estimation of crop residues might be obtained by comparing the residue content of the crop at harvest versus the amount of residue in the soil at planting. Few data exist that can serve as guidelines in this respect. However, our calculations based on the work of Lichtenstein and co-workers[1,2] and Stewart et al.[5] show Rc/Rs (residue in crop/residue in soil) ratios of 0.06 and 0.01 for aldrin and 0.09 and 0.02 for heptachlor in potatoes, respectively. This means that should the tolerance in potatoes be 1 ppm, an acceptable soil residue would vary from 16 to 100 ppm for aldrin and 10 to 50 ppm for heptachlor at these two locations. Lichtenstein's data gave Rc/Rs ratios with carrots of 0.18 and 0.32 for aldrin and heptachlor, respectively. With a 1-ppm food tolerance, soil tolerances would then be 5 and 3 ppm, respectively. These data illustrate variability due to crop and geography (climate and soils). It becomes apparent that a single soil quality value would be either unjust or inadequate, depending upon the level established. Because of interfering compounds, in an economically feasible sense, routine analytically zero quantities of pesticides in soils cannot be determined. Some finite numbers would have to be established for residue tolerances in food crops. Otherwise numbers for a "soil quality index" would be meaningless, and this simple relationship could not be used.

The effect of climatic variability might be well illustrated with some of our own data with the herbicide atrazine. A 3 to 7 year study with oats at three locations in Minnesota indicated that early spring precipitation was a determinant in the appearance of residue injury from atrazine. When only 1 in. of precipitation fell between April 1 and May 15, a residue of 0.6 ppm of atrazine in soil was necessary to sustain an economic yield loss. For each additional inch of rainfall during this period, the amount of atrazine to produce a significant yield reduction was reduced by 0.1 ppm (unpublished data from our laboratory).

In a similar study with soybeans in western Minnesota, rates below 0.5 ppm of atrazine in soil did not produce enough injury to calculate the data as above. A residue of 0.5 ppm produced no damage to soybeans if 4 in. or less precipitation fell between April 1 and June 15. For each inch of precipitation beyond 4 in., 0.5 ppm atrazine residue reduced yields of soybeans by an additional 17%. In south central Minnesota, we found that no significant injury from 0.5 ppm atrazine occurred with up to 12 in. of precipitation during that period (unpublished data).

Differences in residue injury to plants or uptake by plants due to variation in soils is a function of the strength of sorption of the pesticide by soils. Soil factors normally influential in this phenomena are soil pH, organic matter content, and clay content. For example, in our work[3] the LD_{50} of DDT to *Drosophila melanogaster* (contact action) in mineral soils varied from 103 to 790 ppm. In the same soils, the GR_{50} of atrazine to soybeans ranged from 0.16 to 6.30 ppm. At least two orders of magnitude range in the amount of soil residue that can produce injury to crop plants or excessive residues in plants can be expected. Any finite number based on the smallest quantity shown to be a potential hazard would be an extreme disservice and most inequitable because those soils that can tolerate the highest residues are frequently most suitable for production of the crop. For example, vegetable production might be found on irrigated sands or peat soils. With most pesticides, peat soils could accommodate nearly 100 times as much residue as sandy soils without presenting a hazard.

Consequently, for any "soil quality index" for pesticide residues, an equation which includes factors for variations in the crop, climate, soil, and pesticide would be necessary. On a small scale, where one or more of these variables are relatively constant, considerable data are available to make these estimations. Unfortunately, however, data are not available for establishing general equations that could be used by federal control agencies.

REFERENCES

1. Lichtenstein, E. P., G. R. Myrdal, and K. R. Schulz. Adsorption of insecticidal residues from contaminated soils into five carrot varieties, *J. Agr. Food Chem.*, 13:126–131 (1965).
2. Lichtenstein, E. P., and K. R. Schulz. Residues of aldrin and heptachlor in soils and their translocation into various crops, *J. Agr. Food Chem.*, 13:57–63 (1965).
3. Peterson, J. R., R. S. Adams, Jr., and L. K. Cutkomp. Soil properties influencing DDT bioactivity, *Soil Sci. Soc. Amer. Proc.*, 35:72–78 (1971).
4. Smith, D. D., and W. H. Wischmeier. Rainfall erosion, *Advan. Agron.*, 14:109–148 (1962).
5. Stewart, D. K. R., D. Chisholm, and C. J. S. Fox. Insecticide residues in potatoes and soil after consecutive soil treatments of aldrin and heptachlor, *Canad. J. Plant Sci.*, 45:72–78 (1965).

ENVIRONMENTAL INDICES FOR RADIOACTIVITY RELEASES

Paul S. Rohwer and Edward G. Struxness

Health Physics Division
*Oak Ridge National Laboratory**
Oak Ridge, Tennessee 37830

INTRODUCTION

Release of radioactive material to the environment may result in significant radiation exposure of man. Radiation exposure may occur through any one of a number of exposure modes; each mode, in turn, may have any number of subordinate exposure pathways of potential importance. The exposure modes of principal importance following an environmental release of radioactivity may be classified in two groups: (1) internal (radiation source within the body, i.e., inhalation and ingestion) and (2) external (radiation source outside the body, i.e., immersion in contaminated air, submersion in contaminated water, and exposure to a contaminated surface). Adequate assessment of an environmental release of radioactivity requires that consideration be given to possible dose contributions for each of these exposure modes.

THE CUEX CONCEPT

The Cumulative Exposure Index (CUEX) concept is being developed to facilitate realistic assessment of the radiation dose to man as a result of environmental releases of radioactivity. CUEX is defined as a numerical guide indicating the relative significance (dose estimate ÷ dose limit) of measured environmental radioactivity on the basis of the estimated *total dose* to man for *all radionuclides* and *all exposure modes* of importance. The aim in developing this concept is to assess the releases on the basis of time-integrated radionuclide concentrations measured in suitable environmental sampling media; typical measurements would be concentrations in air or water or on the land surface. The measured concentrations are assessed in relation to basic radiation safety standards recommended by recognized authorities for application to members of the public. Because the recommended standards are expressed in units of dose

---

*Operated by Union Carbide Corporation for the U.S. Atomic Energy Commission.

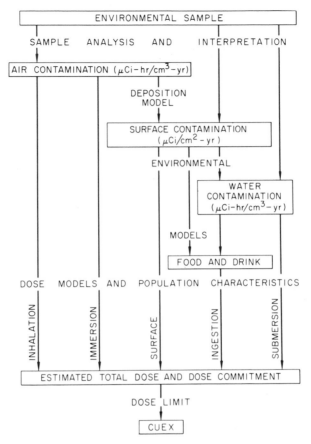

Fig. 1. The CUEX concept.

(rem),* the CUEX concept, of necessity, embodies environmental models and dose models to convert the measured environmental radionuclide concentrations into estimates of radiation dose to man. The final estimate of dose, which includes a contribution for each radionuclide and exposure mode of significance, is compared with the appropriate radiation safety standard to complete the assessment.

The general structure within the CUEX concept is shown in Fig. 1. Starting at the top of the figure, the environmental sample is analyzed and interpreted to estimate the time-integrated radionuclide input to the environment. That input

---

*Throughout this discussion, the term "dose" should be understood to include "dose commitment" whenever internal exposure modes are involved. "Dose commitment" due to an intake of radioactivity is the total dose an individual will accrue within his lifetime as a result of that intake. The intent within the CUEX concept is to assess the total dose to man for a given exposure, not merely a part of that dose (e.g., the dose received in the first year).

is quantified in units of μCi-hr/cm^3-yr from air and water contamination measurements and in units of μCi/cm^2-yr from surface contamination measurements. The integration of environmental input is limited to a period of one year because the recommended radiation safety standards are annual dose limits. Knowledge of the time-integrated radionuclide concentration in air, coupled with dose models and population characteristics, completes the information needed to estimate dose for two exposure modes (radionuclide inhalation and immersion in contaminated air). Dose for the remaining three modes of exposure may be estimated with knowledge of the time-integrated radionuclide concentration measured either in air or water or on the land surface; however, in addition to dose models and population characteristics, there is one other requirement – one that is usually more difficult to satisfy. That requirement is a system of models quantitatively describing radionuclide behavior during the time period between environmental input and man's intake of and/or external exposure to the radioactivity. Summation of the dose contributions from all five exposure modes to obtain an estimate of total dose is shown diagrammatically in the lower portion of Fig. 1. Assessment of an environmental release of radioactivity is achieved by comparing the estimated total dose with the appropriate annual dose limits.

CALCULATION OF CUEX

Environmental releases of radioactivity usually involve a mixture of radionuclides. Therefore, radiological safety assessments of the releases will require estimates of total cumulative doses for a variety of radionuclides and a number of exposure modes. The Cumulative Exposure Index for a given environmental release of radionuclides is calculated in the following manner:

$$CUEX_j = \sum_{i=1}^{n} \frac{E_{ik}}{DLEC_{ijk}}$$

where

$CUEX_j$ = a numerical guide indicating the relative significance (dose estimate $\div$ dose limit) of measured environmental radioactivity on the basis of the estimated total dose in the jth organ for all radionuclides and all exposure modes,

E_{ik} = Time-integrated concentration (μCi-hr/cm^3-yr) for the ith radionuclide in the kth environmental sampling medium, and

$DLEC_{ijk}$ = that time-integrated concentration of the ith radionuclide (μCi-hr/cm^3-yr) which, if present in the kth environmental sampling medium under the conditions considered, is estimated to yield a dose for the jth organ, via all exposure modes, equal to the annual dose limit for that organ.

A detailed presentation of the CUEX development is in press.[4] Calculation of CUEX for a given exposure situation involves a Dose Limit Equivalent Concentration (DLEC) value and an estimate of the time-integrated radionuclide concentration (E) in the environmental sampling medium for each radionuclide present. The units for CUEX, DLEC, and E include two time designations — time-integrated exposure (hours) per unit duration (year) of radionuclide release. Estimation of DLEC's for the individual radionuclides involved is the first step in the CUEX calculation. The annual dose limits used in estimating the DLEC's should be selected by responsible authorities using recommended radiation safety standards. Depending on the situation, the dose limit may be one appropriate for exposure of an individual, or it may be one appropriate for exposure of the population at large. The DLEC's reflect both external and internal dose. Several parameters in the equations used to estimate DLEC's may be varied as a function of the age of the individuals exposed. Therefore, a range of DLEC's, and subsequently CUEX's, may be calculated to cover various age groups in the population. In actual practice it would be prudent to apply the CUEX appropriate for the critical population group. As defined by the International Commission on Radiological Protection (ICRP), "critical population group" refers to a group in the general population whose exposure is homogeneous and typical of that of the most highly exposed individuals in the exposed population.[2]

A $CUEX_j$ value exceeding 1.0 indicates overexposure of the jth organ as result of the cumulative dose contributions for the mixture of radionuclides considered in the concentrations and exposure conditions specified. Selection of the final CUEX may require calculation of several $CUEX_j$ values. The primary reason for multiple calculations is the following: evaluation of an internal exposure involving a mixture of radionuclides is complex because radioactivity accumulation within the body varies among reference organs and among radionuclides, and the result is a spectrum of reference organs irradiated in varying degrees by each radionuclide. For a conservative assessment of the exposure situation, the largest $CUEX_j$ among the reference organs considered would be selected. Adopting the approach used by the ICRP in its recommendations regarding significant exposure of several organs,[3] the exposure of man for a given environmental release of radioactivity would be considered excessive if any one $CUEX_j$ value is greater than 1.0 or if three or more $CUEX_j$ values are each greater than 0.5. The objective in controlling environmental releases of radioactivity on the basis of this concept is to reduce the final CUEX as far below 1.0 as practicable.

APPLICATION OF CUEX

In practice, it will be necessary to calculate DLEC's for each type of release at a given site. This is because the releases will vary from plant to plant at a given site and from site to site, and the environmental factors affecting dispersion and bioaccumulation will be unique to each situation, as will the demographic

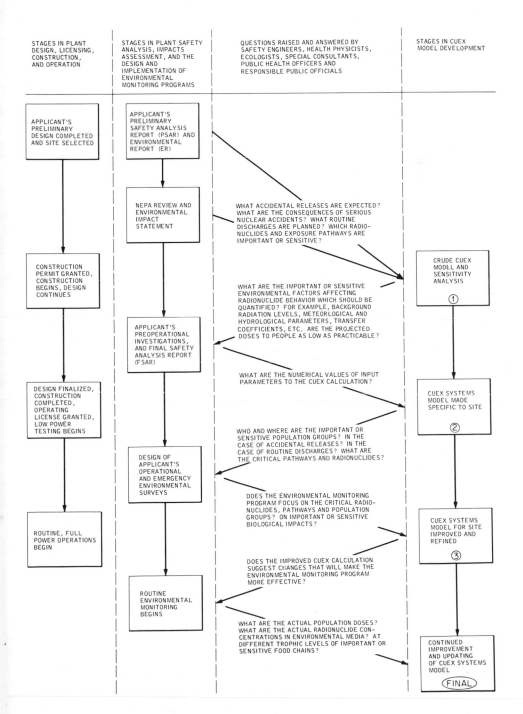

Fig. 2. Modeling and monitoring nuclear plant sites following a systems analysis approach.

factors that affect the doses to people. Thus, a set of DLEC values that might apply generally to most exposure situations is not envisaged here (as, for example, the ICRP and NCRP MPC_a and MPC_w values which are applied to occupational exposure situations). Rather, it is to be expected that health physicists or radiological health specialists, faced with a potential population exposure situation requiring assessment, would calculate and apply their own DLEC's with input data and interpretations that specifically apply to the situation in question.

Figure 2 shows how the CUEX concept could be applied to the safety assessment of radioactive effluents from nuclear plants and how the CUEX can be expected to improve as inputs to the calculation improve. The boxes in column 1 represent significant milestones in siting, designing, constructing, licensing, and operating a power reactor, while those in column 2 indicate some of the important requirements that must be met to qualify, first, for a construction permit and, finally, for an operating license. Arrows to the right in column 3 represent important radiation safety questions that may be raised at those points in time, while the arrows pointing to the left indicate activities and feedback relative to the licensing process and the design of preoperational, operational and emergency surveys. Stages in model development are shown in column 4, indicating how the improvements would evolve in relation to early stages of plant design and construction and continuing on to the point where the plant is operating smoothly and the routine environmental monitoring program appears to be achieving its intended purpose.

A crude systems model could be used early in the first stage (extending that stage back to the time of the applicant's design and site selection) to serve the purposes of the radiation safety analysis in the preliminary safety analysis report (PSAR) and in the environmental impacts statement. The "first cut" assessment would provide a preliminary answer to this question, among many others: What are the consequences of the plant's expected discharges or accidental releases? The information developed and the results obtained in this theoretical analysis of design basis accidents and expected discharges will focus on the gross features of the plant and site that largely affect potential population exposures.

Information on important or sensitive exposure pathways and radionuclides gleaned from the PSAR and the environmental impacts statement can then be used late in stage 1 and in stage 2 to design a systems model specific to the site. These improved inputs to the calculational model will in turn lead to more realistic dose estimates that can serve several purposes: (1) a basis for justifying changes in plant systems design, (2) a basis for designing preoperational surveys, and (3) a basis for assessing radiological safety in the final safety analysis report (FSAR).

Further evolution of the systems model follows from the FSAR through the last part of stage 2 and into stage 3. With information from preoperational studies and the FSAR, together with that developed in the AEC's statement on environmental impact required by the National Environmental Policy Act of 1969, it should be possible to tentatively identify and locate the important or sensitive population groups (i.e., the "critical group" as applied by the ICRP[2] or a "suitable sample" of the exposed population as used by the FRC[1]). In other words, detailed information will be at hand at this stage concerning (1) important radionuclides, (2) important exposure pathways, and (3) important or sensitive population groups to provide a sound basis for designing a realistic dose-estimation systems model. Also, as indicated at stage 2 when the plant is fully licensed and about to become operational the routine environmental monitoring and emergency monitoring systems should have been designed, tested and implemented.

As routine plant operations begin, stage 3, real information on radioactive discharges and actual radiation measurements in the environs will provide the data needed to test the monitoring program for effectiveness and to refine the systems models within CUEX. Such improvements and refinements can be expected to continue (stage 4) as experience is gained and input data accumulated. In this way, the CUEX improves "with age" providing an increasingly meaningful link between the source of radionuclide release (which must be controlled) and the radiation dose to people (which must be kept as low as practicable).

REFERENCES

1. *Background Material for the Development of Radiation Protection Standards*, Federal Radiation Council, Report No. 1, p. 30, para. 5.20. U.S. Government Printing Office, Washington, D.C., May 1960.
2. *Principles of Environmental Monitoring Related to the Handling of Radioactive Materials (Rpt. of Com. 4).* International Commission on Radiological Protection, ICRP Publ. 7, Pergamon Press, London, 1966.
3. *Recommendations of the International Commission on Radiological Protection (Adopted Sept. 17, 1965),* International Commission on Radiological Protection, ICRP Publ. 9, Pergamon Press, London, 1966.
4. Rohwer, P. S., and E. G. Struxness. Development of Radiation Safety Indices for Environmental Releases of Radioactivity, Health Physics Society Annual Meeting, July 11–15, 1971, New York; submitted for publication in *Health Physics* on September 20, 1971.

PLANT INDICATORS IN ECOLOGY*

David F. Grigal

Department of Soil Science and College of Forestry
University of Minnesota
St. Paul, Minnesota 55101

INTRODUCTION

Over five decades ago, Clements[1,2] reviewed the literature dealing with the use of plant indicators in ecology. The result was a volume of nearly 400 pages, with about 450 references. In this definitive work, he stated: "Every plant is a measure of the conditions under which it grows ... an index of soil and climate ... an indicator of the behavior of other plants and of animals in the same spot." Clements considered Hilgard, in 1860, to have reported the first work done with "distinct" indicator objectives. At the rate of growth of scientific literature,[30] a similar review using Clements' perspective should now encompass over 8,000 references in somewhat more than 400 pages. That will not be attempted here. The manipulation of numbers does reflect, however, the volume of literature that is somewhat pertinent to the use of plant indicators in ecological contexts. In fact, much of the current research in terrestrial plant ecology relates directly to the concept of indicators.

CONCEPT

Vegetation – plant life – is a function of its environment, "... a measure of the conditions under which it grows." Jenny[23] has expressed this somewhat more rigorously by the equation

$$v = f(L_0, P_x, t) \tag{1}$$

where v is a property of vegetation, L_0 is the initial state of the ecosystem that includes the vegetation, P_x refers to the external flux potentials that impinge upon the system, and t is the age of the system. L_0, P_x, and t are referred to by Jenny as state factors. The initial state of the system includes such factors as mineralogic composition and topographic configuration, while the fluxes include such factors as precipitation, solar radiation, fire, and floods.

The use of plant indicators in ecology is based on the relationship between some measured property of vegetation and a state factor from functional

*Minnesota Agricultural Experiment Station Scientific Journal Series No. 7859.

equation (1). Where the measured property changes rapidly in relation to the state factor of interest (where the partial differential of the function is at a maximum), good relationships are likely to be found. An extreme example of a condition under which this may be true is on either side of the boundary of a recent fire. Vegetation properties will differ markedly because of the effect of the fire flux. A less extreme example might be the more gradual change in vegetation from prairie to forest with change in climate. An example where change is occurring very slowly, on the other hand, might be the change in vegetation associated with a small change in mean annual temperature. In the temperate zone, this change is usually difficult to detect. The primary requisite in the use of indicators, then, is the selection of a vegetation property that changes rapidly with change in the state factor of interest. Most state factors investigated, except in the most basic research, are associated with man's economic interests, such as the presence of agronomically productive conditions or exploitable mineral resources. In the former case, interest is not directed toward any single state factor, but instead is directed toward a combination of those factors that produce a demonstrable effect.

The value of indicators is related to the passage of time. In some studies, the primary goal may simply be determination of time or age of the system based on a property of vegetation. In dendrochronology, crossdating of growth rings of old trees with those in beams in buildings is used to determine dates of occupancy of archaeologically important sites.[16] In other studies, however, the passage of time may mitigate the effect of a state factor. In quantitative terms, we have moved to a region where the partial differential of equation (1) has decreased. An example of this mitigation might be in the effects of a fire as discussed earlier. Such effects will be most noticeable immediately after passage of the fire front, while in time they will become less and less noticeable. After 100 years, only a few indications of the fire may remain, and after 500 years, other state factors will likely have exerted a stronger influence so that the effects will have disappeared. This dampening influence of time, which must be considered in all indicator studies, is an aspect of ecological succession. This basic principle of ecology is the orderly process of (plant) community change toward a final mature community called the climax.

INDICATORS

As stated earlier, the property of vegetation that has the maximum indicator value for a given state factor is that which is most strongly affected by a change in the factor. Choice of a property that accomplishes this has elicited considerable controversy in the literature. Resolution of this controversy can best be left to the participants, but it may be useful to note that as different systems are studied, the most useful vegetation property may also differ. Three broad categories of vegetation properties are used as indicators: (1) floristic composition, (2) morphology, and (3) elemental composition.

Floristic Composition

A classic plant indicator is a single species whose presence or absence indicates the level of a state factor. For example, in Rhodesia, basil (*Ocimum homblei*) does not grow on soil containing less than 100 ppm copper.[8] Its presence, therefore, has led to the discovery of several ore deposits. Generally, however, it has been recognized[12,33] that the entire plant community, or at least a large part of it, should be used as an indicator of conditions. The community carries more redundant, but therefore more reliable, information about the state factors. Controversy exists, however, over whether subordinate or dominant species in the community should be used. The controversy is centered about the countering arguments that the dominant species are most important in terms of community processes and should therefore be used as indicators[12] versus the proposition that subordinate species are often more exacting in their requirements than are dominants and should therefore more accurately reflect the level of a state factor.[29]

Morphology

Morphology, usually used as an index of the physiological state (vigor) of plants, has been used to indicate the level of state factors. In areas where high levels of potentially toxic minerals occur, the toxicity often results in changes in morphology and plant vigor.[8] Levels of elements not generally considered to be toxic may also affect the vigor of plants whose requirements and tolerance differ significantly from the levels in the soil. In some areas of Russia, larch (*Larix dahurica* Turcz.) grow poorly on soil high in carbonates and can therefore be used to map the presence of carbonate rocks.[50]

Not only may a specific state factor be assessed by the morphology of an indicator plant or plants, but the entire state factor complex may also be assessed. For example, work in both western and northern United States has shown that the growth and vigor of bracken fern (*Pteridium aquilinum* L.) may be a good indicator of the productivity of a site for forest trees.[18,27]

Elemental Composition

If the goal of the use of plant indicators is to identify a state factor such as a mineral deposit, then levels of that mineral within the plant tissue can often be used as an index of its levels in the substrate. For example, Lyon and Brooks[25] have investigated the elemental content of tree daisy (*Olearia rani* A. Cunn.) and have found that this species was especially valuable for biogeochemical prospecting for molybdenum.

LIMITATIONS

As has been outlined thus far, and will be expanded upon later, the use of plant indicators in an ecological context is widespread and well accepted. Care must be taken, however, when applying indicator significance to the floristic composition of plant communities or the morphology or elemental composition of individual plants. This caution is necessary because of the presence of ecotypes or "distinct races resulting from the selective action of a particular environment and showing adaptation to that environment."[49] Ecotypes can only be identified by comparing their growth in a uniform environment. Some of the earliest work demonstrating ecotypic variation was reported by Clausen, Keck, and Hiesey,[11] who found races of plants along an altitudinal gradient. Ecotypes may differ in response to state factors, and this alters the very basis of the indicator concept. Workman and West[47] studied the value of a single species as an indicator. They used winterfat (*Eurotia lanata* (Pursh) Moq.), a species traditionally regarded as a dependable indicator of soil conditions (e.g., Shantz[34]). Using both germination studies and transplant tests, they found relatively wide ecotypic variation in tolerance to soluble salts. They recommended that cognizance of this variation be taken into account when the species is used to indicate soil conditions. In England, some plant species have been found to evolve a tolerance to toxic heavy metals in a relatively short time (50 to 100 years).[6] At least in some cases, then, as state factors change, plants will change with them.

On a community basis, McMillan[26] showed that the joint occurrence of the same series of taxa at different sites is not sufficient evidence for considering the habitats to be similar. He studied five species of grasses from across Nebraska. Under uniform conditions, there was a distinct difference in the morphology of individuals of all species at any time which was related to the location from which they had been obtained.

Finally, use of elemental composition of a plant species as an indicator of elemental content of soil can also be affected by ecotypic variation. Steinbeck[36] found that Scotch pine (*Pinus sylvestris* L.) grown from seed obtained from 122 stands over the natural range of that species and outplanted at three different uniform sites showed significant differences in elemental composition related to the site at which they were grown. This tends to verify the concept of indicators. Unfortunately, he also found significant differences in elemental composition at each site which were related to the place of origin of the seed, tending to negate the concept of indicators.

These studies indicate the danger of broadly applying the results of a localized study of plant indicators. The valid use of indicators requires observation and experimentation wherever they are to be used. Perhaps this point is best made by the results of three unrelated studies of the effect of grazing pressure on species composition of grasslands. Blue grama (*Bouteloua gracilis* (H.B.K.) Lag.) occurs widely in the western United States. In New Mexico,[31] the proportion of blue grama was significantly lower on grazed versus ungrazed sites. In

Colorado,[22] grazing appeared to have no effect on the proportion of blue grama, while in Oklahoma,[21] blue grama increased with grazing. None of these studies is necessarily incorrect. They show the extent of variability in species response, due both to ecotypic variation and to unmeasured environmental or state factors.

EXAMPLES

The remainder of this paper will briefly discuss a few examples of the use of plant indicators in specific contexts. The types of uses described and the specific works cited are not meant to be inclusive. As pointed out in the introduction, literally thousands of specific examples of the use of indicators can be found by an imaginative mind in a well-stocked library. Clements[12] classified indicators into four broad types: (1) condition or factor indicators, reflecting environmental factors such as light, temperature, etc.; (2) process indicators, indicating natural or artificial disturbance; (3) practice or use indicators, suggesting the possibility of a cultural practice such as agriculture or forestry; and finally (4) paleic indicators, extending the use of indicators into the geologic past by the use of fossilized remains. That classification will be used here in order to give some correspondence to his work. However, overlap is inevitable when examples of each category are discussed. Also, based on Jenny's equation of state,[23] the definition of some of the categories will be altered.

Condition indicators

Condition indicators can be considered to be those that reflect state factors associated with the initial state of the system (and its subsequent states over time) in equation (1). State factors that can be assessed include mineralogy of the substrate, water table depth, and topographic configuration. The use of condition indicators is extensive, and considerable interest and effort has been expended on this use in the Soviet Union (e.g., Chikishev[10]), especially as the vast, sparsely inhabited regions such as Siberia undergo settlement and development.

Classic examples in plant ecology of condition indicators are those associated with serpentine soils. Serpentine areas have been widely studied botanically because of their lack of productivity, their unusual floras, and the distinctive physiognomy of the vegetation that is present.[45] Analyses of serpentine soils have shown them to be low in Ca and high in Mg, Cr, and Ni,[42] and the distinctive vegetation is considered to be a reflection of the soil properties. Because of the response of plants to mineral levels, botanical prospecting is being done in many areas.[8] Whitehead and Brooks[44] used not only the level of U in vegetation to determine U levels in the soil, but carrying the concept a step further, they used levels of α radioactivity in the vegetation as a determinant of areas potentially valuable for exploitation. This α radiation was not associated primarily with U but with other radioactive elements such as Ra and Po.

Water is also a state factor whose relative level can be ascertained by the use of indicators. Minore[28] used the morphology of skunk-cabbage (*Lysichitum americanum* Holt and St. John) to indicate water table depth. Zedler and Zedler[51] found the distribution of a number of species correlated with microtopographic differences in water table depth. Soil moisture levels themselves have also been related to vegetation distribution.[14] One of the classical uses of indicators has been in the determination of levels of salinity in soils.[12] This use has continued, with emphasis being placed on studies in local areas so that salinity can be determined with relatively high accuracy.[15,38,39]

Process Indicators

Clements[12] considered indicators to be useful in identifying such processes as grazing, fire, and floods. In the context of the state factor equation (1), these would be considered to be fluxes that impinge upon the system. Though Clements included temperature, light, and climate as conditions, not processes, these too are fluxes and should also be considered in the context of process indicators.

Plants certainly indicate climatic differences. Major plant community types, such as tropical rain forests or temperate grasslands, are closely associated with given climates,[46] as their names indicate. Generally, the size of the dominant plant species tends to decrease as dryer and cooler climatic conditions prevail. On a smaller scale, differences in plant communities also occur on slopes with different aspects,[9] because each slope is subjected to a different climate due to differences in solar flux.

Fire has a profound effect on the plant community, and this effect can be observed for literally hundreds of years after a fire has passed. Some species of plants are adapted to the occurrence of fire. Jack pine (*Pinus banksiana* Lamb.), for example, has cones that protect viable seed from damage by fire. After a fire, seeds are released to a seedbed on which conditions are often ideal for germination and growth.[1] Other species, such as trembling aspen (*Populus tremuloides* Michx.), prolifically sprout from roots following a fire.[13] Presence of extensive communities of these species is evidence for the prior occurrence of fire.[1] This evidence remains until the species disappear from the stand through succession, a process that may take hundreds of years.

Grazing is another flux that has been classically determined by the use of indicator plants. The degree of grazing is shown by two types of indicators, native species that may decrease or increase in response to grazing and those species that invade because of disturbance associated with grazing.[43] The species that decrease are usually those preferred by the animals and those physiologically vulnerable to grazing. Increases usually occur in less preferred species. Invading species are often annuals, and although there are exceptions, most invaders are also of low desirability. The fundamental relationships among species in their response to grazing are relatively well-known, and range condition classes are often assigned to areas by managing agencies on the basis of those species present.[37] The degree of use allowed by the agencies is determined by the condition classes. Clements[12] discussed some grazing indicators at length in his book, and these plant—process relationships continue to be explored. The earlier cited works with blue grama[21,22,31] are examples of such studies.

Practice Indicators

When vegetation properties are used to indicate the potential for cultural practices, the properties are actually being used as an index of the entire state factor complex in equation (1). Cultural practices usually involve growing plants (crops) on the site, and thus, indicators are used whose response to the state factor complex is similar to that of the cultured species.

Clements[12] and Shantz[33] devoted considerable discussion to indicators of land with potential agronomic value. Though that concept remains valid, many of their examples have only heuristic value. Most of the land once occupied by indicators of good agricultural productivity in the United States is now in crop production, and land value is measured by crop yield rather than estimated by indicators. In developing countries, however, the use of indicators is still a valid aid to land settlement. In East Africa, vegetation and landform are surveyed to assess potential productivity of grazing land, and productivity categories are then used in land management.[3,32]

Indicators have been extensively used for forestry purposes,[20] and at least one book has been published on the general subject.[2] Cajander[7] was one of the first foresters to use indicators systematically, and he used plant associations to classify forest types for management purposes. He defined these types by the vegetation on the site when the stand was normally stocked and mature, and such a definition was possible without reference to the species of trees on the site. This classification of forest stands proved useful in Finland, where it was developed, both for identifying potential yield and for providing a basis for selection of silvicultural treatment. Though the system has been tried elsewhere (e.g., Heimburger[17] in the United States and Sisam[35] in Canada), it has never achieved great popularity, perhaps because the other attempts have been made in areas floristically more complex than Finland. Work has continued, however, on other applications of indicators in forestry. For example, both Vallee and Lowry,[40] studying black spruce (*Picea mariana* (Mill.) B.S.P.), and Hodgkins,[19] studying longleaf pine (*Pinus palustris* Mill.), have shown that indicator plants can be used to reasonably estimate the potential productivity of a site.

Paleic Indicators

The final category of indicators discussed by Clements[12] is really an extension of the use of indicators into the past, usually based on fossilized remains of plants. An unusual use of indicators that relates to the past is in archaeology, where areas of past human activity can often be recognized by the characteristic vegetation growing on them.[4] This is due to the fertilizer value of refuse from human habitation (such as bones), which enhances plant growth, and to the variety of species brought to the site for food or medicinal purposes.

The major use of fossilized remains of plants as indicators has been in the area of stratigraphic pollen analysis. Pollen grains and spores from vascular plants can accumulate and be preserved for long periods of time in lake sediments. Sequences of such pollen in the sediment can then be used to indicate change in plant composition of an area over time. Such changes are then related to presumed climatic changes in the past. For example, pollen and plant macrofossils indicate that 6000 to 8000 years ago prairie vegetation extended into east-central Minnesota. Before that period, and at present, the dominant vegetation in the area was mixed conifer—deciduous forests. This vegetation history has been used as evidence for a pronounced xerothermic climatic interval.[48] The presence of mineralized plant remains in soils[41] also has been used to indicate the extent of prairie vegetation. Studies using remains of plants to infer past climates must necessarily presume similarity in response of vegetation then and now.

CONCLUSIONS

Plant indicators have been, and will continue to be, extensively used in ecological contexts. Technological advances, such as the use of sophisticated data processing equipment to relate plant distribution to environmental factors[5] and the introduction of remote sensing to detect subtle differences in vegetation over vast areas,[24] have helped further refine the application of the indicator concept. It is likely that plant indicators will be used in the future for purposes never dreamed of by Clements.

REFERENCES

1. Ahlgren, I. F., and C. E. Ahlgren. Ecological effects of forest fires, *Bot. Rev.*, 26: 483–533 (1960).
2. Aichinger, E. *Pflanzen als forstliche Standortsanzeiger* (Plants as indicators of forest sites), Osterreichischer Agrarverlag, Vienna, 1967; cited in *For. Abstr.*, 29:3293 (1968).
3. Astle, W. L., R. Webster, and C. J. Lawrance. Land classification for management planning in the Luangwa Valley of Zambia, *J. Appl. Ecol.*, 6: 143–169 (1969).
4. Bank, T. P., II. Ecology of prehistoric Aleutian village sites, *Ecology*, 34: 246–264 (1953).
5. Barkham, J. P., and J. M. Norris. Multivariate procedures in an investigation of vegetation and soil relationships of two beech woodlands, Cotswold Hills, England, *Ecology,* 51: 630–639 (1970).
6. Bradshaw, A. D., T. S. McNeilly, and R. P. Gregory. Industrialization, evolution and the development of heavy metal tolerance in plants, *British Ecological Society Symposium*, 5: 327–343 (1965).

7. Cajander, A. J. The theory of forest types, *Acta Forest. Fenn.*, 29: 1–108 (1926).

8. Cannon, H. L. Botanical prospecting for ore deposits, *Science*, 132: 591–598 (1960).

9. Cantlon, J. E. Vegetation and microclimates on north and south slopes of Cushetunk Mountain, New Jersey, *Ecol. Monogr.*, 23: 241–270 (1953).

10. Chikishev, A. G., (ed.). *Plant indicators of soils, rocks, and subsurface waters.* Consultants Bureau, New York, 1965.

11. Clausen, J., D. D. Keck, and W. M. Hiesey. *Experimental studies on the nature of species. I. Effect of varied environments on western North American plants,* Carnegie Inst. Wash., Publ. 520, 1940.

12. Clements, F. E. *Plant indicators.* Carnegie Inst. Wash., Publ. 290, 1920.

13. Daubenmire, R. F. *Plants and environment* (2nd Ed.). John Wiley and Sons, New York, 1959.

14. Daubenmire, R. F. Soil moisture in relation to vegetation distribution in the mountains of northern Idaho, *Ecology*, 49: 431–438 (1968).

15. Dodd, J. D., and R. T. Coupland. Vegetation of saline areas in Saskatchewan, *Ecology*, 47: 958–968 (1966).

16. Glock, W. S. Tree growth. II. Growth rings and climate, *Bot. Rev.,* 21: 73–188 (1955).

17. Heimburger, C. C. Forest type studies in the Adirondack Region, Cornell University Agr. Exp. Sta. Memoir 165, 1934.

18. Hellum, A. K. Factors influencing frond size of bracken on sandy soils in northern Lower Michigan, Ph.D. thesis, University of Michigan, 1964; *Dissert. Abstr.,* 28: 2207B (1967).

19. Hodgkins, E. J. Productivity estimation by means of plant indicators in the longleaf pine forests of Alabama, pp. 461–474 in *Tree growth and forest soils.* C. T. Youngberg and C. B. Davey (eds.). Oregon State Univ. Press, Corvallis, 1970.

20. Hustich, I., chairman. Symposium on forest types and forest ecosystems, *Silva Fennica*, 105: 1–142 (1960).

21. Hutchinson, G. P., R. K. Anderson, and J. J. Crockett. Changes in species composition of grassland communities in response to grazing intensity, *Proc. Okla. Acad. Sci. 1966*, 47: 25–27 (1968).

22. Hyder, D. N., R. E. Bement, E. E. Remmenga, and C. Terwilliger, Jr. Vegetation-soils and vegetation grazing relations from frequency data, *J. Range Manage.*, 19: 11–17 (1966).

23. Jenny, H. Derivation of state factor equations of soils and ecosystems, *Soil Sci. Soc. Amer. Proc.,* 25: 385–388 (1961).

24. Johnson, P. L., (ed.). *Remote sensing in ecology.* Univ. Georgia Press, Athens, 1969.

25. Lyon, G. L., and R. R. Brooks. The trace element content of *Olearia rani* and its application to biogeochemical prospecting, *New Zealand J. Science,* 12: 200–206 (1969).

26. McMillan, C. Nature of the plant community. I. Uniform garden and light period studies of five grass taxa in Nebraska, *Ecology*, 37: 330–340 (1956).

27. Minore, D. The growth of *Pteridium aquilinum* (L.) Kuhn var. *pubescens* in relation to geographic source, soil nutrients, and forest site, Ph.D. thesis, University of California, Berkeley, 1966; *Dissert. Abstr.*, 27: 705B (1966).

28. Minore, D. Yellow skunk-cabbage (*Lysichitum americanum* Holt and St. John) – an indicator of water-table depth, *Ecology*, 50: 737–739 (1969).

29. Odum, E. P. *Fundamentals of ecology* (3rd Ed.). W. B. Saunders, Philadelphia, 1971.

30. Passman, S. *Scientific and technical communication.* Pergamon Press, New York, 1969.

31. Pieper, R. D. Comparison of vegetation on grazed and ungrazed pinyon-juniper grassland sites in southcentral New Mexico, *J. Range Manage.*, 21: 51–53 (1968).

32. Pratt, D. J., P. J. Greenway, and M. D. Gwynne. A classification of East African rangeland, *J. Appl. Ecol.*, 3: 369–382 (1966).

33. Shantz, H. L. Natural vegetation as an indicator of the capabilities of land for crop production in the Great Plains area. USDA, Bureau of Plant Industry Bull. 201, 1911.

34. Shantz, H. L. Plants as soil indicators, pp. 835–860 in *Soils and Men*. USDA, Yearbook of Agriculture, 1938.

35. Sisam, J. W. B. Site as a factor in silviculture. Canada Dept. Mines and Resources, Silv. Res. Note 54, 1938.

36. Steinbeck, K. Site, height, and mineral nutrient content relations of Scotch pine provenances, *Silvae Genetica*, 15: 42–50 (1966).

37. Stoddart, L. A., and A. D. Smith. *Range management* (2nd Ed.). McGraw-Hill, New York, 1955.

38. Ungar, I. A. Species-soil relationships on the Great Salt Plains of northern Oklahoma, *Amer. Midl. Natur.*, 80: 392–406 (1968).

39. Ungar, I. A., W. Hogan, and M. McClelland. Plant communities of saline soils at Lincoln, Nebraska, *Amer. Midl. Natur.*, 82: 564–577 (1969).

40. Vallee, G., and G. L. Lowry. Forest soil-site studies. II. The use of forest vegetation for evaluating site fertility of black spruce. Pulp and Paper Research Institute of Canada, Woodland Papers 16, 1970.

41. Verma, S. D., and R. H. Rust. Observations on opal phytoliths in a soil biosequence in southeastern Minnesota, *Soil Sci. Soc. Amer. Proc.*, 33:749–751 (1969).

42. Walker, R. B. The ecology of serpentine soils. II. Factors affecting plant growth on serpentine soils, *Ecology*, 35: 259–266 (1954).

43. Weaver, J. E., and F. E. Clements. *Plant ecology* (2nd Ed.). McGraw-Hill, New York, 1938.

44. Whitehead, N. E., and R. R. Brooks. Radioecological observations on plants of the Lower Buller Gorge region of New Zealand and their significance for biogeochemical prospecting, *J. Appl. Ecol.*, 6: 301–310 (1969).

45. Whittaker, R. H. The ecology of serpentine soils, *Ecology*, 35: 258–259 (1954).
46. Whittaker, R. H. *Communities and ecosystems.* Macmillan, New York, 1970.
47. Workman, J. P., and N. E. West. Ecotypic variation of *Eurotia lanata* populations in Utah, *Bot. Gaz.*, 130: 26–35 (1969).
48. Wright, H. E., Jr., and W. A. Watts. Glacial and vegetational history of northeastern Minnesota. University of Minnesota, Minnesota Geological Survey Special Publication SP–11, 1969.
49. Wright, J. W. *Genetics of forest tree improvement.* United Nations FAO, Forestry and Forest Products Studies 16, 1962.
50. Zagrebina, N. L. The relationship between vegetation and rock lithology in the Daldynsk region of the Yakutian ASSR from aerial photographs, pp. 120–123 in *Plant indicators of soils, rocks, and subsurface waters.* A. G. Chikishev (ed.). Consultants Bureau, New York, 1965.
51. Zedler, J. B., and P. H. Zedler. Association of species and their relationship to microtopography within old fields, *Ecology*, 50: 432–442 (1969).

INDEX

A

Academy of Natural Sciences,
  Philadelphia 93
Adrenal cortex 215–217
Agricultural Stabilization and
  Conservation Service 159
Air pollutants (*see specific materials*)
Air quality indices 9–11, 19,
  183, 199
  Coefficient of Haze 190, 192
  Fulton County (Georgia) Index 187
  National Wildlife Federation's
    EQ Index 9–11
  Oak Ridge Air Quality Index
    (ORAQI) 189–196
  Ontario Index 190
  PINDEX 187–190, 194–195
  Plants as indicators (*see also*
    Plants) 101
  Soiling Index 192
Algae 93–95
American Association for the
  Advancement of Science v, 13
Animal tissues (*see also specific*
  tissues) 84–89, 112–114, 117,
  215–219, 222–229
Antimony 85
Antioch College 43
Aristotle iii
Arsenic 85
Asbestos 84
Atlanta, Georgia 187
Atomic Energy Commission 1, 8,
  109, 183, 199, 249

B

Bacteria 93, 117
Barnard College 55
Battelle Memorial Institute 71
Beltsville, Maryland 103
Bernard of Chartes 211
Beryllium 85, 115–116
Biochemical indicators (*see also*
  Animal tissues *and* Enzymes) 2, 109
Biochemical oxygen demand 175,
  178–179
Biological indicators (*see also*
  Animal tissues, Enzymes, Plants,
  and specific classes of animals) 2,
  83, 93, 101, 109, 255
Blood 84–89, 112–113, 218,
  225–229
Broken-stick model 99
Brown University 133
Bureau of Land Management 159
Bureau of Mines 166–167
Bureau of Outdoor Recreation
  149–151

C

Cadmium 16, 84–85, 115
California Department of Water
  Resources 180
Carbon dioxide 79–80
Carbon monoxide 24–25, 79–80,
  184–189, 191, 193, 201–203
Cell cultures 114
Cell-free systems 115
Census Bureau 160